The Little Mathematical Decision Formula

To James —
An institution at Umstead already! Hope you enjoy!
Best Regards,
[illegible]

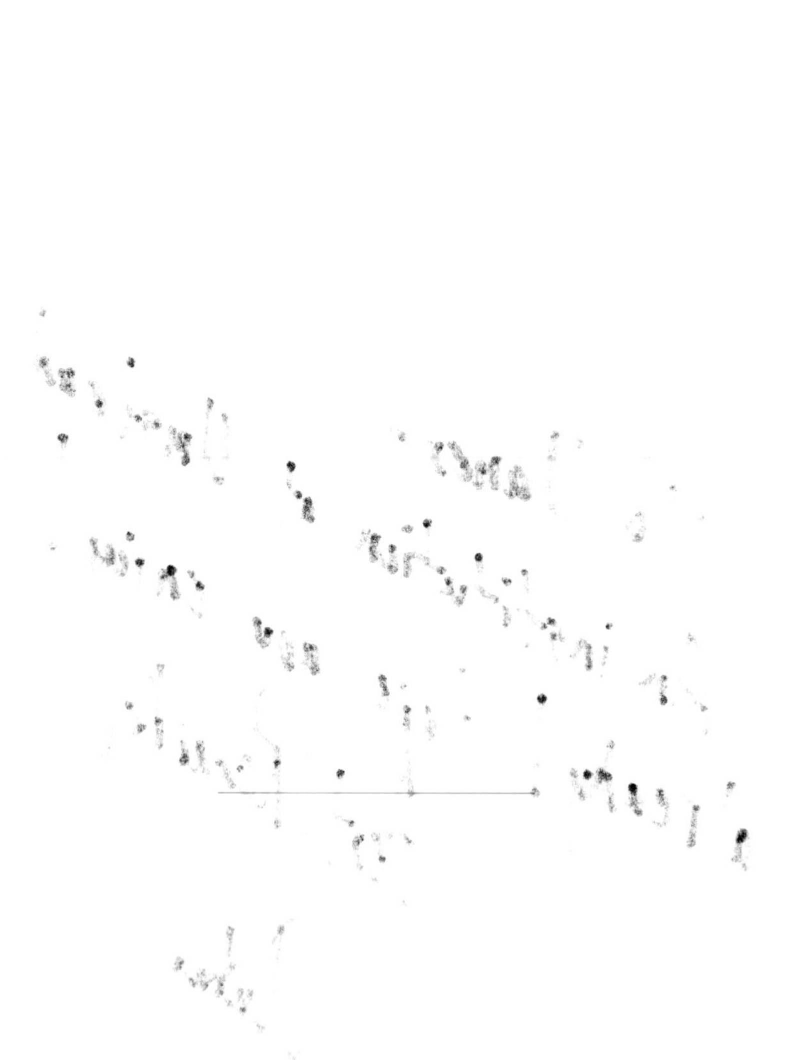

The Little Mathematical Decision Formula

•••

How to use simple math for life's not-so-simple decisions

Andrew Ferrari

ISBN: 1499578504
ISBN 13: 9781499578508
Library of Congress Control Number: 2014909289
CreateSpace Independent Publishing Platform
North Charleston, South Carolina

Contents

Introduction

Decisions and choices are central to life, and a big reason life can be so difficult and confounding. We decide when to go to bed, when to get up, what to eat, what to wear, where to go shopping, who to hang out with. We decide where to live, where to work, what car to drive, what apartment to rent, what house to buy, what college to attend. We decide who to befriend, who to date, who to marry. In our modern-day millennial lives especially, where there are myriads of options for everything, decisions and choices are a continual requirement. Without choices life would certainly be simpler—but it would also be less interesting and fulfilling.

The importance of decision-making is obvious. The course and quality of our lives depend almost completely on the decisions we make, sometimes in small ways, but other times in huge, life-altering ways that can influence health, happiness, and even survival. Our decisions can bring us fulfillment and joy, but they can also bring us unhappiness, misery, and regret.

Making decisions can be a difficult, taxing affair, especially when a decision is both complex and critical. Consequently we have a variety of ways to help ourselves through the process. We consult family and friends. We confer with professional counselors. We consider the pluses and minuses; we weigh the pros and cons. We appeal to our instincts and intuition. And any or all of the above! In the end, we somehow push through, and for the most part we do all right. Nonetheless, erroneous decisions seem an inescapable part of the human condition.

Mathematics can help. In these pages is presented a simple mathematical model that offers an easy, reliable tool for making decisions in many situations in life. At the heart of the model is a very simple formula—the *Little Mathematical Decision Formula*, a.k.a. the *LMDF*—which involves only the elementary operations of addition, subtraction, multiplication, and division. One merely answers a few questions about a decision in a numerical fashion, plugs these numbers into the formula, and performs a simple computation, from which the resulting decision is clearly indicated. The beauty of this formula is that although decisions are produced in a mathematically precise way, they are still very much one's own, based on one's desires, tastes, values, and priorities. It is a remarkable little formula, and can be a powerful tool for making decisions in life.

The *LMDF* is applicable to many decisions, but not all of them. Indeed, as we'll see, there are certain decisions for which the formula is difficult or impossible to implement, and for which blatantly erroneous decisions will be produced. However, for those decisions to which the formula properly applies, the *LMDF* will yield eminently sound decisions. Not infallible decisions to be taken as gospel, but sound decisions that can and should be taken seriously in one's decision process.

This book consists of five chapters. *Chapter 1* introduces the *Little Mathematical Decision Formula* and describes in general terms how it models decision-making. *Chapter 2* then takes a closer look and reveals exactly how the formula produces its sound decisions. *Chapter 3* constitutes the practical core of the book and presents a simple four-step procedure for using the *LMDF* to make decisions. *Chapter 4* then demonstrates how this procedure can be applied to six typical life decisions. Finally, *Chapter 5* rounds out the picture by offering a brief look at a few alternative decision models that might be of interest and use to the reader.

With a little time and effort anyone can learn the *Little Mathematical Decision Formula* and thus acquire a useful tool for making decisions in life. In the process, one will attain a greater understanding of decision-making, as well as a greater understanding and appreciation of the power of mathematics to resolve some of life's more difficult challenges. Best wishes for all of your decision-making endeavors!

1

The Little Mathematical Decision Formula

Suppose you're buying a new car and trying to decide between models. Now, if you've only one criterion for your decision, say price, then the decision is easy—you simply choose the car with the lowest price.

Almost surely, however, you'll have multiple criteria—say price, appearance, power, handling, and gas mileage—and your decision won't be so simple. Now, if one car happens to be superior in *all* these respects, then the decision will still be easy. Usually, however, this will not be the case, and one car will be better in some respects but worse in others, and you will have to somehow weigh these competing factors and identify the car that is best for you *overall.* As we know, such decisions are not always easy.

And if that weren't enough, oftentimes you'll also prioritize your criteria, placing more importance on one than another. For example, you might feel that gas mileage is more important than power and handling. This kind of prioritization adds even more complexity to a decision, and makes it that much more difficult to sort things out and identify your best car.

This type of decision is quite typical in life, and not just for deciding on a car. We summarize its four basic elements. First, there are certain criteria or *metrics* on which a decision will be based. Second, there is a prioritization or *weighting* of these metrics according to their respective levels of importance. Third, there are *evaluations* of the alternatives with respect to the metrics. Finally, there is a *decision calculus* by which all this information is processed and the best alternative is identified:

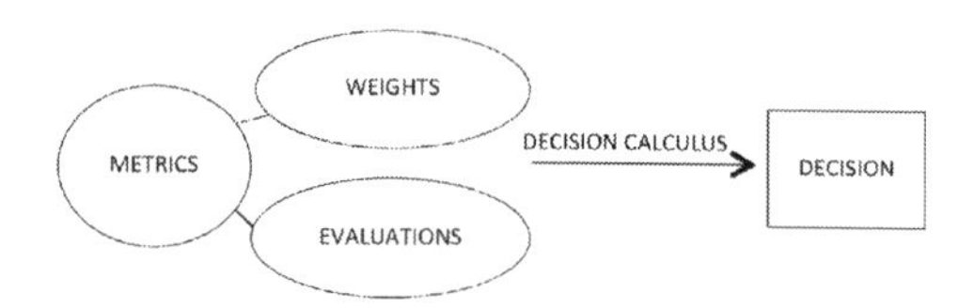

This same decision process is used by us humans for a great many decisions in life regardless of what we might be deciding on—a phone, computer, car, restaurant, house, job, college, or whatever.

The *Little Mathematical Decision Formula* is a precise mathematical representation—or *model*—of this human decision process. Now, in order that such a formula be possible in the first place, the elements of this decision process must be translated into precise mathematics, that is, the process must be *modelled.*

The first step in accomplishing this is that certain elements of the decision process must be *quantified* so they can be represented *numerically* in a mathematical formula. There are two such elements—the *weightings* of the metrics and the *evaluations* of the options with respect to these metrics. Now, I won't say right now specifically how this quantification is to be accomplished—this is discussed in *Chapter 3*—but I will say how the quantities will appear in the formula.

To keep things simple for right now, I'll illustrate with the car example above. Suppose we've chosen three metrics for this decision—price, power, and appearance—and that we've determined their respective weightings and have evaluated the cars with respect to them. First, these weightings are quantified by assigning to each metric a numerical *weight*, denoted by w_1, w_2, and w_3. Likewise, these evaluations are quantified by assigning numerical evaluations, denoted by v_1, v_2, and v_3. These w's and v's are then exactly the quantities that will appear in the *LMDF*—precise quantified versions of the weightings and evaluations that are fundamental to human decision-making.

All that remains to complete our model is that the decision calculus employed to ultimately arrive at a decision—namely our processing of our evaluations of the cars together with our weightings of the metrics—must somehow be represented in a mathematically precise way. Again, I won't say right now specifically how this will be accomplished—this is discussed in *Chapter 2*—but the end result is exactly the *LMDF*, which will indeed take the numerical versions of our weightings and evaluations—namely the w's and v's—and combine them mathematically in just the right way so as to yield a decision. It will do so by producing a precise numerical score for each car, the highest of which will represent the best car for us overall given this choice of metrics, weights, and evaluations.

Here, then, is the *Little Mathematical Decision Formula* as it applies to this example; later we'll discuss it in its full generality. The letter S stands for the score for a car:

$$S = \frac{w_1 v_1 + w_2 v_2 + w_3 v_3}{w_1 + w_2 + w_3}$$

Don't be frightened by this formula! It might look like a handful, but it is in truth extremely simple mathematically. Indeed, by means of only the elementary operations of addition, subtraction, multiplication, and division, it takes the three numerical weights and three numerical evaluations and produces a score for each car. No exponents or logarithms, no trigonometric functions, and certainly no calculus—just *arithmetic*.

The formula, however, *looks* complicated because of the gaggle of letters and subscripts. The only purpose of the letters, though, is to represent the weights and evaluations *generally*, so they could stand for whatever numerical values might be chosen for the decision. And the only purpose of the subscripts is to differentiate between the weights and evaluations corresponding to the three different metrics. In actual practice, the formula will ultimately involve only *numbers*—the specific numerical weights and evaluations chosen—such as, for example, the following:

$$S = \frac{3 \cdot 7 + 5 \cdot 2 + 9 \cdot 5}{7 + 2 + 5}$$

The simplicity of the formula is now evident.

Looking at the general version of the formula above, we can see how closely it models our human decision process. Indeed, we can see that it takes the *w*'s and *v*'s—the quantified versions of our weightings of the metrics and our evaluations of the cars—and by means of a precise mathematical calculation produces a score for each car and thereby a decision:

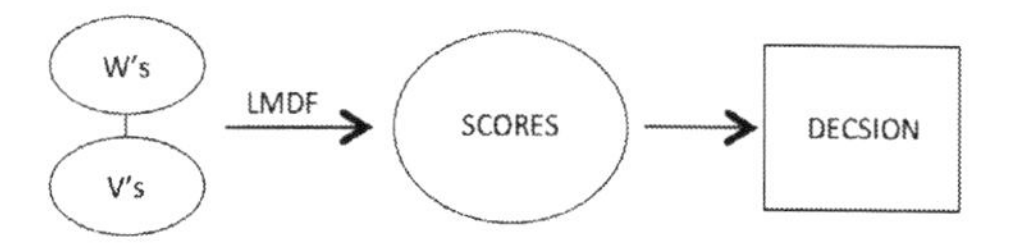

The close parallel with the decision process can clearly be seen.

We can moreover see exactly how the *LMDF* works mathematically. The numerator—which turns out to be the real crux of the formula—consists of multiplying together each pair of weights and evaluations and then adding these products together. The denominator—which plays only a minor role—consists merely of summing the three weights. The results from the numerator and denominator are then divided, producing a score. This calculation can generally be performed in a matter of seconds.

The formula is exactly the same for every car being considered. Now, the actual *numerical values* that will be *plugged into* the formula will generally be different for different cars, but the formulas themselves will be the same. Actually, the weights w_1, w_2, w_3 will be identical for every car—since the same metrics are used for each—but the evaluations v_1, v_2, v_3 will usually be different for different cars. In any case, the end result will be scores for each car overall, the highest of which will represent the best car for this choice of metrics, weights, and evaluations. And it doesn't matter how many cars are involved—two or two hundred. Exactly the same formula is used for each and every one of them.

Generally speaking, the same essential formula is utilized for any decision to which this model applies, no matter what the alternatives nor how many there might be. The only aspect of the formula that changes is the *number* of terms on the top and bottom, which merely represents the number of metrics utilized for a decision. If there are five such metrics, there will be five terms on the top and bottom. If there are 10 metrics, there will be 10 terms. And if there are 100, there will be 100. Here, for example, is the formula for a decision involving five metrics:

$$S = \frac{w_1 v_1 + w_2 v_2 + w_3 v_3 + w_4 v_4 + w_5 v_5}{w_1 + w_2 + w_3 + w_4 + w_5}$$

As you can see, aside from the increased number of terms on the top and bottom, the formula is essentially the same. Now, the *computation* obviously becomes lengthier as more metrics are utilized and a greater number of alternatives are considered, but the essence of the formula doesn't change. Consequently, no matter what the decision nor how many alternatives or metrics are involved, the formula will still take the corresponding weights and evaluations, just as in our car example, and combine them in the same way to produce scores for each alterative and thereby a decision.

At this point we've become fairly well acquainted with the *Little Mathematical Decision Formula* and have attained a fairly good understanding of how it models our human decision process in a mathematically precise way.

This understanding, however, is incomplete. In the next chapter, therefore, we take a closer look at the *LMDF* and the scores it produces, examining in particular what these scores really mean and why they constitute a sound basis for making decisions. Such understanding will be important in *Chapter 3* where we'll use the formula to make actual decisions.

2

How Formula Works

By now we understand many things about decision-making in general and the *Little Mathematical Decision Formula* in particular. We understand that most decisions involve four basic elements regardless of the alternatives being considered, namely, the criteria or metrics on which a decision is based, the prioritization or weighting of these metrics, the evaluation of alternatives with respect to these metrics, and finally the decision calculus by which all this information is processed and the best alternative is identified.

We also understand something about how the *Little Mathematical Decision Formula* corresponds to this human decision process. Here, for reference, is the formula for a decision involving three metrics:

$$\frac{w_1 v_1 + w_2 v_2 + w_3 v_3}{w_1 + w_2 + w_3}$$

We understand that the w's represent numerical versions of our weightings of the metrics, and that the v's represent numerical versions of our evaluations of the alternatives with respect to these metrics. Moreover, we understand that the formula itself represents a mathematically precise version of the decision calculus by which we process this information and arrive at a decision. Specifically, we understand that the formula mathematically combines the numerical weights and evaluations in just the right way so as to a produce precise scores for each alternative, and hence a decision as to which is best overall given this choice of inputs.

What we don't yet fully understand, however, is this last part—how exactly the formula produces meaningful scores for the alternatives. This is the real mystery of the formula. What do these scores really mean, and why do they constitute a sound basis on which to make decisions?

To answer this question, let us take a step back and pretend that we aren't yet aware of the formula, but that we are trying to *create* such a formula, that is, create a way to mathematically combine the v's and w's so as to produce meaningful scores for the alternatives.

Reflecting on this, a first idea might be simply to compute the *average* of the evaluations for each alternative. For instance, supposing there are three metrics for a decision, the score for each alternative would then be given by the following:

$$\frac{v_1 + v_2 + v_3}{3}$$

The best alternative would then simply be the one with the highest such average.

At first glance, this might seem a simple and reasonable way to mathematically compute scores and determine a best alternative, however it turns out to be profoundly flawed and will in fact yield the wrong decision in almost every situation.

A simple example makes this clear. Suppose we are deciding between two breakfast cereals solely on the basis of the metrics of *taste* and *nutritional value*. Now suppose that one cereal—call it *Cereal 1*—tastes great but is not particularly healthy, and is thus given numerical evaluations of +8 for taste and -8 for nutritional value. Here we are using a scale from -10 to +10 for the evaluations, where the negatives represent bad quality and the positives good, with -10 being the worst and +10 the best. Suppose that the other cereal—call it *Cereal 2*—represents the exact opposite of *Cereal 1*, with a not-so-great taste of -8 but a great nutritional value of +8. In other words, each cereal is as good in one respect as it is bad in the other. Now, if we compute the average of the evaluations for each cereal, in both cases the +8 and -8 will exactly cancel each other out to produce scores of 0. Therefore neither cereal should be preferable.

For most people, however, one cereal *would* be preferable. The reason is that for most people either taste or nutritional value would naturally take priority and would thus be weighted more heavily in the decision process. Those for whom taste takes priority would prefer the great-tasting *Cereal 1*, and those for whom nutrition is more important would prefer the healthful *Cereal 2*. Merely averaging the evaluations for each cereal, therefore, for most people, has not produced the right scores or the right decision.

The critical flaw with this kind of average, generally speaking, is that it completely ignores any prioritization of decision criteria, which, as we've seen, is almost always a factor in human decision-making. Of course, this is obvious already from the averaging formula itself, which involves only evaluations but not weights. Such an average, then, is an exceedingly poor model of decision-making and is doomed from the start to produce incorrect scores and decisions in almost every situation. The only exception is the rare decision for which all metrics are deemed to have equal importance.

What is required, therefore, for a sound decision model is a formula for averaging evaluations that somehow incorporates differing levels of importance of metrics. But how could this be accomplished? A crucial hint, it turns out, is provided by the failed averaging formula itself, but in order to see this we must first take a closer look at why exactly it failed.

To this end, we recall that in calculating the scores for *Cereal 1* and *Cereal 2* via an average, the +8 and -8 evaluations for taste and nutritional value exactly cancelled themselves out, producing scores of 0 for both cereals:

$$+8 \quad + \quad -8 \quad = \quad 0$$

In reality, however, as we saw, the scores for most people should *not* both be 0, but one score should be larger than the other, because most people would prioritize either taste or nutritional value and hence prefer one cereal over the other.

Another way of looking at this is that for most people the +8 and -8 evaluations for taste and nutrition should *not* cancel each other out for both cereals. But how could this ever happen? To answer this, we suppose for definiteness that nutritional value is prioritized over taste, in which case the score for the healthy *Cereal 2* should come out *greater* than the score for the not-so-healthy *Cereal 1.*

We consider *Cereal 1.* Now, if its -8 evaluation for nutritional value, due to the greater importance of this metric, could somehow count *more* than its +8 evaluation for taste, then a score *less* than 0 would be produced:

$$\scriptstyle +8 \quad + \quad \displaystyle \Large -8$$

Likewise, for *Cereal 2*, if its +8 evaluation for nutritional value, due to the greater importance of this metric, could somehow count more than its -8 evaluation for taste, then a score *greater* than 0 would be produced:

$$\scriptstyle -8 \quad \displaystyle + \quad \Large +8$$

The result of all of this is that *Cereal 2* would indeed have a higher score than *Cereal 1*, and would thus be identified as the better cereal, as it should.

Mathematically, of course, this is complete nonsense, but it has nonetheless served a valuable purpose—to point us toward what kind of averaging formula will generally be needed to incorporate a prioritization of metrics and hence be a sound decision model. Such a formula, instead of counting all evaluations *equally* in the averaging process as does a normal average, must somehow count as *greater* the evaluations corresponding to the more important metrics, and count as *less* the evaluations corresponding to the less important metrics.

It turns out, in fact, that there is exactly such a formula. It represents a more sophisticated kind of averaging process called a *weighted average*—and it is exactly what the *Little Mathematical Decision Formula* is!

Finally, having now come full circle, we can begin to answer our original question and understand what the scores produced by the *LMDF* really mean, and why exactly they constitute a sound basis for making decisions. Here again, for reference, is the formula for a decision involving three metrics:

$$\frac{w_1 v_1 + w_2 v_2 + w_3 v_3}{w_1 + w_2 + w_3}$$

We note first how the evaluations v_1, v_2, and v_3 in the numerator no longer stand alone as they did in a simple average, but how each is now multiplied by a weight w. This has exactly the necessary effect of giving greater or lesser relative numerical value to each v according to the size of the w by which it is multiplied, this size corresponding exactly to the degree of importance of the corresponding metric. The result is that in the final score the more important evaluations will make a proportionately greater numerical contribution, and the less important ones a proportionately lesser contribution. This is in direct contrast to a simple average, in which all evaluations contribute equally. Such a score therefore represents a much truer and more accurate valuation of an alternative overall, and provided the metrics, weights, and evaluations were chosen properly, it constitutes a much sounder basis on which to make a decision.

We've ignored so far the denominator $w_1 + w_2 + w_3$ because, despite its appearance, it doesn't actually contribute in any substantive way to the final scores or decision. Indeed, since it is identical for every alternative, its presence merely causes the respective scores to be scaled down uniformly, leaving the ordering of the scores, and hence the final decision, unaffected. In mathematical parlance, the denominator is merely a *normalization factor*. What it does accomplish, however, just as the denominator of a simple average, is to ensure that all the scores fall on the same numerical scale as the separate evaluations—a nice feature when one wants to know the overall quality of an alternative in this absolute sense.

We can see more precisely what is going on with a weighted average if we go back to the previous example. There we considered *Cereal 1*, whose taste and nutritional value were evaluated respectively with an excellent +8 and a dismal -8, and *Cereal 2*, whose evaluations represented the exact reverse. Supposing here again that nutritional value is more important than taste, we now choose suitable numerical weights. Using a scale from 0 to 10—zero representing no importance and 10 extreme importance—we select weights of 5 for taste and 9 for nutritional value. The *LMDF* then yields the following scores for *Cereals 1* and *2*:

$$C_1 = \frac{5(+8) + 9(-8)}{9+5} = -\frac{32}{14} \qquad C_2 = \frac{5(-8) + 9(+8)}{9+5} = \frac{32}{14}$$

The better cereal has therefore been properly identified as *Cereal 2*.

From these formulas we can see precisely what is going on with the weighted averages. For *Cereal 1*, the +8 evaluation for taste in the numerator has been multiplied by the weight of 5, but the -8 evaluation for nutritional value has been multiplied by the larger weight of 9 because of its greater importance. This causes this more important -8 evaluation to make a proportionately greater contribution to the final score than the +8 evaluation:

$$5(+8) \quad + \quad \mathbf{\Large 9}(-8)$$

The score has thereby been correctly driven down to the negative value indicated.

For *Cereal 2*, on the other hand, the -8 evaluation for taste has been multiplied by the weight of 5, but the +8 evaluation for nutritional value has been multiplied by the larger weight of 9 because of its greater importance. This causes this more important +8 evaluation to make a proportionately greater contribution to the final score than the -8 evaluation:

$$5(-8) \quad + \quad \mathbf{\Large 9}(+8)$$

The score has thereby been correctly driven up to the positive value indicated.

The resulting decision, again, is that *Cereal 2* is the better choice given this prioritization of metrics. The denominator of 14, as one can see, does not play any substantive role in the ordering of the scores or the final decision.

The effect of the weights actually becomes even clearer if we rewrite the formulas slightly:

$$C_1 = \frac{5}{14}(+8) + \frac{9}{14}(-8) \qquad C_2 = \frac{5}{14}(-8) + \frac{9}{14}(+8)$$

Here we have simply moved the denominator of each formula under each weight in the numerator, which leaves the value of the scores unchanged. The weights now appear in their *normalized* form, namely $\frac{5}{14}$ and $\frac{9}{14}$.

First we note that the weights in this form are fractions that sum to 1. This means, in effect, that they represent *percentages* which together total 100%. The significance of this is that now their effect is crystal clear. Indeed, we see that each weight, via its multiplication with an evaluation, causes exactly the right fraction or percentage of that evaluation to contribute to the final score according to its exact level of importance. The more important a metric, the larger will be the percentage contribution to the final score of the evaluation with respect to that metric; the less important, the smaller the contribution. And such weighted averages work in exactly the same way regardless of what decision is being made or which metrics, weights, or evaluations are being employed.

Finally we have a complete answer to our original question of how exactly the *Little Mathematical Decision Formula* works, that is, what its scores really mean and why they constitute such a sound basis for making decisions:

> The *Little Mathematical Decision Formula* is a weighted average of the evaluations of an alternative, whose weights represent the exact levels of importance of the metrics on which these evaluations are based. In this way, exactly the right fraction or percentage of each evaluation contributes to the final scores in exact proportion to its level of importance. In the end, therefore, the scores produced by the *LMDF*—provided the metrics, weights, and evaluations were chosen properly—represent true and accurate overall valuations of each alternative in accordance with a decision-maker's desires, tastes, judgments, values, and priorities, and hence they constitute a sound basis for making decisions.

Our understanding of the *LMDF* is now complete, and we are thus fully prepared for the practical task of actually utilizing it to make decisions—the subject of the next chapter.

3

Making Decisions: A Four-Step Procedure

Armed now with a thorough understanding of the *Little Mathematical Decision Formula*, we are prepared to utilize it to make actual decisions. There are four steps:

Step 1: Choose Metrics
Step 2: Assign Weights to Metrics
Step 3: Evaluate Options with Respect to Metrics
Step 4: Compute Scores & Identify Best Option

These steps represent the four basic elements of human decision-making as they manifest themselves in the framework of our mathematical model. The first three steps represent *our* part in the process—providing the basic ingredients of a decision and expressing them numerically. The last represents the *LMDF*'s part—processing this numerical data and producing scores and ultimately a decision. In this chapter we show how to carry out these steps, providing guidelines and hints along the way for doing so.

Before proceeding, however, two remarks must be made regarding the decisions that can and cannot be handled by the *LMDF*. First, the scores produced by the *LMDF* can be utilized in a variety of ways, for a variety of different decisions. Here are three basic types:

- *Identifying Best Alternative:* This is perhaps the most common type of decision in life, and the primary focus in this book. Employing the *LMDF*, one only has to identify the alternative with the highest score.

- *Short-Listing Alternatives:* One can also use the *LMDF* to short-list alternatives, such as when one is trying to identify the particular jobs or colleges to which to apply, or the particular cars to test drive. Here one only has to identify the alternatives whose scores one judges to be sufficiently high. Naturally, the *LMDF* can then be applied a second time to identify the best of these short-listed alternatives, perhaps with a more discriminating set of metrics.

- *Ranking Alternatives:* Finally, the scores produced by the *LMDF* can obviously be used to rank alternatives from best to worst.

In the end, the scores produced by the *LMDF* offer much flexibility, and can be utilized however one sees fit.

Finally, though the *LMDF* applies to many decisions in life, it does not apply to *all* of them. There are two principal kinds for which it does *not* apply. Here we describe them generally, and in the sections that follow we take a closer look:

- *No Decisions Involving Extraordinary Factors.* The *LMDF* does not apply to decisions involving factors of extraordinary character or importance, such as decisions directly affecting one's health, safety, or survival. It is strictly for decisions of a more ordinary nature.

- *No Decisions Involving Unknown or Uncertainty.* The *LMDF* does not apply to decisions involving factors that are unknown or uncertain. It is strictly for decisions for which all the necessary ingredients are known and definite.

Do not attempt to apply the *LMDF* to the wrong kind of decision! As we'll see, for such decisions the formula will be difficult or impossible to properly implement and will almost certainly produce unsound decisions. Other than these decisions, however, most any other decision in life is suitable for this model, and if carried out properly will produce eminently sound decisions that can and should be taken seriously in one's decision process.

Step 1: Choose Metrics

Fundamental to effective decision-making is the choice of the right yardsticks or criteria by which alternatives are measured and compared—the right *metrics.* Such metrics are the foundation of the entire decision process, and the soundness of the final scores and decision depend completely on their choice.

To carry out this step, reflect carefully on your options and identify all the factors associated with them that have any importance or relevance to you for the decision at hand. These criteria could be of a practical, emotional, or even a spiritual nature—it doesn't matter at all. Any factor whatsoever that is relevant to you for the decision should be included. This requires, above all, that you know yourself and know your options, and that you be honest about what really does and doesn't matter to you for a decision.

Below are a few basic guidelines for choosing metrics that will help ensure the best set of metrics and the best decision possible from the *LMDF*:

- *Include All Relevant Metrics:* It is important to include *all* metrics that are relevant to you for a decision, and not to omit any, for otherwise the scores produced by the *LMDF* are likely to be incorrect and the resulting decision unsound.

- *Minimize Number of Metrics:* Try at the same time, for the sake of simplicity, to choose the *minimum* number of metrics. Omit any that are superfluous or redundant.

- *Seek Independent Metrics:* For the simplest, cleanest decision with the least margin for error, endeavor to choose metrics that are as independent from one another as possible, that is, those which have as little in common, or as little dependence on one another, as possible. This will minimize redundancies and help ensure the most accurate scores and soundest decisions.

Here's a typical example. Suppose we're trying to decide between two restaurants for a special birthday dinner involving a few close friends and family. We choose our metrics, endeavoring to identify all the relevant ones, to choose the minimum number possible, and to choose those that are as independent from one another as possible. Suppose we choose the following six:

Price Range
Food Quality
Service Quality
Liveliness
Proximity to Home
Attractiveness

Other metrics might have been chosen as well, such as a restaurant's location, setting, size, popularity, prestige, or clientele, but the six above are the ones we've deemed relevant for this decision.

The choice of metrics generally, of course, is a very individual matter, depending completely on one's objectives, priorities, and experiences, as well as the circumstances surrounding a decision. If, for example, our birthday dinner were just for ourselves and a significant other, we might include a metric for how romantic a restaurant is, but might omit the metrics of liveliness and proximity to home. On the other hand, if the dinner were just an everyday meal with family or friends, we might simply choose the metrics of food quality, price range, and proximity to home. Someone wealthy might exclude the metric of price range altogether. And someone preferring quiet might omit the metric of liveliness. Whatever the case, the important thing in the end is that one choose metrics that represent what is relevant *to them* for a decision.

In the end, this step will usually pose few difficulties, but for a certain type of decision, identifying all the relevant metrics—and *knowing* one has done so—could be difficult or impossible. Such decisions were discussed generally at the beginning of the chapter, but now we can be more specific:

- *No Decisions with Relevant Information Missing.* If a decision involves options for which information is missing, such blind spots could cause one to unwittingly omit important metrics, resulting in an incomplete set of metrics and an incorrect decision. Going back to our example, if one of the restaurants, unbeknownst to us, will be hosting on the night of our dinner a live band that plays loudly and badly, then our lack of a

> metric regarding a restaurant's music might cause us to wrongly choose this restaurant on the basis of the other metrics. One must know one's options well!

We caution the reader to beware of decisions of this type, because the *LMDF* may not be able to be properly implemented and may therefore produce unsound decisions.

Once suitable metrics have been chosen, the *LMDF* can be set up for each option. Here are the formulas for *Restaurant 1* and *Restaurant 2*, which contain six terms on the top and bottom corresponding to the six metrics:

$$R_1 = \frac{w_1 v_1 + w_2 v_2 + w_3 v_3 + w_4 v_4 + w_5 v_5 + w_6 v_6}{w_1 + w_2 + w_3 + w_4 + w_5 + w_6}$$

$$R_2 = \frac{w_1 v_1 + w_2 v_2 + w_3 v_3 + w_4 v_4 + w_5 v_5 + w_6 v_6}{w_1 + w_2 + w_3 + w_4 + w_5 + w_6}$$

Step 2: Assign Weights to Metrics

Fundamental also to effective decision-making is the assignment of the proper degrees of importance to the chosen metrics—the proper *weights*. This step is of critical importance because it establishes the correct prioritization of the metrics—essential to obtaining accurate scores and a sound decision in the end.

To carry out this step, reflect carefully on each metric and ask yourself how important it is to you for the decision at hand. Now, being in the context of a mathematical model, this level of importance must be *quantified*, that is, it must be expressed as a *numerical value* along some kind of numerical scale. Here we use a simple scale from 0 to 10, where 0 represents a metric of no importance and 10 represents one of extreme importance:

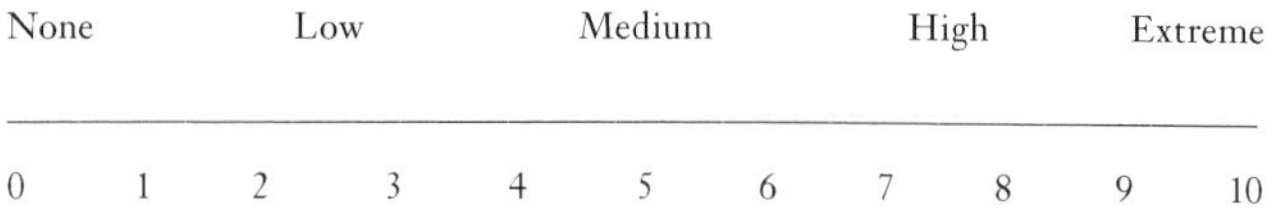

None Low Medium High Extreme

0 1 2 3 4 5 6 7 8 9 10

Level of Importance of a Metric

For each metric, then, simply choose a number along this scale that best represents its degree of importance to you for the decision at hand. The most important thing is that these values be as true and accurate as possible, for the closer they match the actual importance levels of the metrics, the truer and more accurate will be the final scores according to your values and priorities, and the sounder will be the resulting decision. Insofar as these weights are inaccurate, however, the final scores will be disconnected from you personally and will lack true meaning, and any decision based on them will be unreliable. Such accuracy requires, above all, that you know yourself and are honest about how much or little something truly matters to you for a decision. Absolute precision, however, is not possible since the human judgments, opinions, and feelings underlying these numerical choices cannot be measured perfectly.

To illustrate, we return to the example from *Step 1*, where we were deciding between two restaurants for a special birthday dinner. There we chose six metrics, and now we assign them weights, endeavoring to match these weights as closely as possible to the true level of importance we judge each metric to have:

Price Range	*3*	*Liveliness*	*7*
Food Quality	*8*	*Proximity to Home*	*3*
Service Quality	*8*	*Attractiveness*	*8*

As these numbers indicate, food quality, service quality, and attractiveness are very important factors to us for this decision, and liveliness is not far behind. Of relatively low importance to us are price range and proximity to home.

Naturally, just as with metrics, the choice of weights in general is a very individual matter, depending completely on one's objectives, priorities, and experiences, as well as the circumstances surrounding a decision. If, for example, our dinner were just an everyday meal with family or friends, then price range and proximity to home might take precedence and be weighed more heavily than above, but food quality, service quality, and attractiveness might be weighed less heavily. Someone frugal would likely weigh much more heavily the metrics of price range and proximity to home, but weigh much more lightly, or perhaps omit altogether, the other metrics. A foodie would no doubt weigh food quality even more heavily than we did. Whatever the case, the important thing in the end is that one assign weights that accurately reflect the importance levels of the metrics *to them.*

There are two ways to check the accuracy of weights if one should feel so inclined:

- *Check Ordering of Weights:* Check that the numerical ordering of the weights from largest to smallest matches your actual prioritization of the corresponding metrics from most important to least. If they match, great; if not, the weights should be modified accordingly.

- *Check Relative Value of Weights:* Check that the relative numerical values of the weights match the actual relative importance of the corresponding metrics. For example, if you judge one metric to be much more important than another, then its numerical weight should be correspondingly larger. But if one metric is of equal or only slightly differing importance than another, then their numerical weights should be correspondingly close in value. If such comparisons match, great; if not, the weights should be adjusted accordingly.

We illustrate these checks with our restaurant example, beginning with the first method. Examining the numerical ordering of our chosen weights, we see that the most important metrics have been chosen to be food quality, service quality, and attractiveness, followed by liveliness, and ending with price range and proximity to home. As this indeed matches our actual prioritization of the metrics, the weights check out.

Moving to the second method, we examine the numerical weights and see, first, that all three metrics of food quality, service quality, and attractiveness were assigned a weight of 8, which indeed matches our judgment that all three have about the same importance. Moreover, we see that the metric of liveliness was assigned a slightly lower weight of 7, which matches our feeling that it is only slightly less important than those other three metrics. Finally, the substantially smaller weight of 3 assigned to price range and proximity matches our judgment that these two metrics are considerably less important than the others. Once again, our weights check out.

In the end, this step will usually pose few problems, but for a certain type of decision it will be difficult or impossible to properly carry out. Such decisions were discussed generally at the beginning of the chapter, but now we can be more specific:

- *No Decisions with Metrics of Extraordinary Importance:* If a decision involves any metric having extraordinary importance that far exceeds that of the other metrics—off-the-scale importance as it were—then it will be difficult or impossible to assign it the proper weight on the same 0-to-10 scale as the other metrics.

For example, if one were deciding on a doctor to perform a critical surgery, one might include the metric of *friendliness* and assign it a weight of 7, but what weight then should be assigned to the metric of *competence*? The maximum allowed by our scale is 10, but even this weight would hardly reflect the true importance of this metric, especially in relation to the 7 assigned to friendliness. For most of us, the importance of competence would so far outweigh the importance of friendliness that it would be impossible to assign it a suitable weight on the same 0-to-10 scale; rather, it should be assigned a much larger weight, say 50, 100, or even 1000. The *LMDF*, therefore, would not be applicable.

This example is not unique. There are a number of decisions in life like it, principally those involving health, safety, or survival. We caution the reader, therefore, to beware of such decisions, because for them the *LMDF* will likely be difficult or impossible to correctly implement, and any decision produced by it will almost certainly be unsound. So don't even attempt it! Such decisions must be handled by other means. In *Chapter 5* we take a brief look at an alternative model that is better suited for such decisions, though still not without substantial difficulties to properly implement.

Once the appropriate weights have been selected, they can be entered into the *LMDF* for each option in whatever sequence one pleases. The formulas for *Restaurant 1* and *Restaurant 2* then become:

$$R_1 = \frac{3v_1 + 8v_2 + 8v_3 + 7v_4 + 3v_5 + 8v_6}{3 + 8 + 8 + 7 + 3 + 8}$$

$$R_2 = \frac{3v_1 + 8v_2 + 8v_3 + 7v_4 + 3v_5 + 8v_6}{3 + 8 + 8 + 7 + 3 + 8}$$

Step 3: Evaluate Options with Respect to Metrics

Fundamental also to effective decision-making is the true and accurate evaluation of each alternative with respect to the chosen metrics. Such evaluations are of critical importance for obtaining accurate scores and a sound decision in the end.

To carry out this step, reflect carefully on each option and for each metric ask yourself how good or bad that option is with respect to that metric in your judgment. Now, again, being in the context of a mathematical model, this level of quality must be quantified and thus expressed as a numerical value along some kind of numerical scale, as were the weights. Here, however, such a scale must entail both good and bad quality, and therefore we use a scale from -10 to +10, where the positive values represent good quality, the negatives bad quality, and zero neither good nor bad:

Extremely Bad	Very Bad	Fairly Bad	Slightly Bad	Neutral	Slightly Good	Fairly Good	Very Good	Extremely Good

-10 -8 -6 -4 -2 0 +2 +4 +6 +8 +10

Quality of an Option With Respect to a Metric

For each option, then, consider each metric and simply choose a number along this scale that best represents its level of quality with respect to that metric in your judgment. The most important thing is that these values be as true and accurate as possible, for the closer they match the actual quality levels of the options, the truer and more accurate will be the final scores according to your judgments and tastes, and the sounder will be the resulting decision. Insofar as these evaluations are inaccurate, however, the final scores will be disconnected from you personally and will lack true meaning, and any decision based on them will be unreliable. Such accuracy requires, above all, that you know yourself and are honest about how good or bad an option truly is to you. Absolute precision, however, is not possible since the human judgments, opinions, and feelings underlying these numerical choices cannot be measured perfectly.

Don't be at all concerned with the weights at this point, and don't take them into account in your evaluations—they will be incorporated automatically later on. Simply evaluate each option with respect to each metric *independent* of the degree of importance of the metric.

To illustrate, we return to the previous example, where we were deciding between two restaurants for a special birthday dinner. We had already chosen six metrics and assigned them weights, and now we evaluate each restaurant with respect to those metrics, endeavoring to match these evaluations as closely as possible to the true level of quality we judge each restaurant to have:

	Restaurant 1	*Restaurant 2*
Price Range	+5	-5
Food Quality	+7	+9
Service Quality	+7	+9
Liveliness	+8	+2
Proximity	+8	+2
Attractiveness	+7	+8

As these numbers indicate, *Restaurant 1* for us is very good for its liveliness and proximity to home; nearly as good for its food quality, service quality, and attractiveness; and only fair for its price range. *Restaurant 2* we've judged to be extremely good for its food and service quality, nearly as good for its attractiveness, and only slightly good for its liveliness and proximity to home. Its price range for us is fairly bad.

Naturally, just as with metrics and weights, the evaluation of options in general is a very individual matter, depending completely on one's tastes and preferences, as well as the circumstances of a decision. Experienced restaurant-goers with very high standards, for example, might assign lower marks than us for food and service quality for both restaurants, whereas a less finicky customer might assign even higher ones. Someone wealthy might feel that the price range of *Restaurant 2* is fine and give it much higher marks than we did, whereas someone in financial difficulty might evaluate it even more negatively. A lively, popular restaurant will be exciting and stimulating for some, but annoying and bothersome for others. And what is close-by for one will be far-away for another. Whatever the case, the important thing in the end is that one's evaluations reflect accurately the level of quality of each option *for them*.

There are two ways to check the accuracy of evaluations if one should feel so inclined:

- *Check Ordering of Evaluations:* For each option, check that the numerical ordering of its evaluations matches your actual ordering of its quality levels with respect to the various metrics. If they match, great; if not, the evaluations should be modified accordingly.

- *Check Relative Value of Evaluations:* This can be done in two ways. First, evaluations of the *same* option can be compared for different metrics. Namely, for each option, check that the relative numerical value of its evaluations match its actual relative quality with respect to the corresponding metrics. For example, if you judge the quality of an option with respect to one metric to be much higher than with respect to another, then the corresponding numerical evaluation should be suitably larger. But if the quality of an option with respect to different metrics is equal or only slightly different, then the corresponding evaluations should be close or equal in value. Moving to the second way to check the relative value of evaluations, one can compare them *across different options* for the *same metric.* Namely, check that the relative numerical value of evaluations of different options with respect to the same metric match their actual relative quality for that metric. For example, if you judge the quality of one option with respect to some metric to be much higher than the quality of another option for the same metric, of if you judge them to be of equal or only slightly differing quality, then this should be reflected in the corresponding numerical evaluations.

We illustrate these checks with our restaurant example, beginning with the first method. Examining the numerical ordering of our evaluations for *Restaurant 1*, we see that its best qualities have been chosen to be its liveliness and proximity to home, followed by its food quality, service quality, and attractiveness, and ending with its price range. As this indeed matches our actual ordering of these qualities, the evaluations for *Restaurant 1* check out. A similar check can be performed for *Restaurant 2*.

Moving to the second method, we begin by checking the relative values of the evaluations for each restaurant separately. Examining the evaluations for *Restaurant 2*, we see, first, that both its food and service were given the same evaluation of +9, which indeed matches our judgment that both are of equally high quality. Moreover, we see that its attractiveness was given a slightly lower evaluation of +8, which matches our feeling that this quality is just slightly below that of its food and service. Finally, the substantially lower evaluations of +2, +2, and -5 given to its proximity, liveliness, and price range match our judgment that these qualities are considerably less to our liking than the other qualities above. Once again, our evaluations for *Restaurant 2* check out, and a similar check can be performed for *Restaurant 1.*

Finally, we check the relative value of the evaluations of each restaurant with respect to the same metric. For example, looking at the metric of price range, we see that our evaluations of +5 and -5 for the two restaurants are greatly disparate, which indeed matches our judgment that one is much more affordable than the other. For the metric of attractiveness, on the other hand, we see that the evaluations of +7 and +8 are very close, which matches our feeling that one restaurant is only a little more attractive than the other. Similar comparisons can be performed for other metrics. Again, our evaluations check out.

In the end, this step will usually pose few real difficulties, but for certain types of decisions it will be difficult or impossible to carry out properly. Such decisions were discussed generally at the beginning of the chapter, but now we can be more specific:

- *No Decisions with Options of Extraordinary Quality:* If a decision involves any option having extraordinarily good or bad quality with respect to a metric that far exceeds either its quality with respect to other metrics or the quality of another option—off-the-scale quality as it were—then it will be difficult or impossible to assign it the proper evaluation on the same -10 to +10 scale as the other evaluations. For example, if one were considering an attempt of the summit of Mt. Everest, the tallest mountain in the world, then what would be a suitable evaluation of this endeavor with respect to the metric of *safety*? For most of us, the danger of such a venture would be so extraordinary that it would be impossible to assign it a suitable evaluation on the same -10 to +10 scale as more ordinary evaluations such as for cost or travel time; rather, it should be assigned a much more negative value, say -50, -100, or even -1000. The *LMDF*, therefore, would not be applicable.

- *No Decisions with Options of Unknown or Uncertain Quality:* If a decision involves any option having unknown or uncertain quality with respect to any metric, such an option will obviously be impossible to properly evaluate. For example, if the food quality at a particular restaurant is unknown or inconsistent, then a definitive, accurate, and reliable evaluation would be impossible. The *LMDF* therefore does not apply.

We caution the reader to beware of such decisions, because for them the *LMDF* will likely be difficult or impossible to correctly implement, and any decision produced by it will almost certainly be unsound. So don't even attempt it! Such decisions must be handled by other means. In *Chapter 5* we take a brief look at alternative models that are better suited for such decisions, though still not without substantial difficulties to properly implement.

Once the appropriate evaluations have been determined, they can be entered into the *LMDF* for each option—but each evaluation must be correctly paired with the weight corresponding to the same metric! The formulas for *Restaurant 1* and *Restaurant 2* then become:

$$R_1 = \frac{3(5) + 8(7) + 8(7) + 7(8) + 3(8) + 8(7)}{3 + 8 + 8 + 7 + 3 + 8}$$

$$R_2 = \frac{3(-5) + 8(9) + 8(9) + 7(2) + 3(2) + 8(8)}{3 + 8 + 8 + 7 + 3 + 8}$$

Step 4: Compute Scores & Identify Best Option

Finally, once metrics have been chosen, weights have been assigned, and evaluations have been performed, all that remains is that this information be processed in just the right way so that a sound decision is produced—and this is exactly what the *Little Mathematical Decision Formula* does!

Simply plug the numerical weights and evaluations into the suitable version of the *LMDF* for each option—the version whose numerator and denominator contain the same number of terms as the number of metrics—and carefully compute the final scores. This last step is the easiest of all, the culmination of all the good work of the previous steps.

Once scores have been computed, they can be utilized either to identify the best option, to short-list options, to rank them from best to worst—or for whatever other purpose one sees fit. Scores very close in value, however—say within a point—should be considered a tie due to the slight imprecision inherent in assigning numerical values to human judgments, opinions, and feelings as required by *Steps 2* and *3*. Such too-close decisions should be made by other means.

For our restaurant example, the computation of the scores goes as follows:

$$R_1 = \frac{3(5) + 8(7) + 8(7) + 7(8) + 3(8) + 8(7)}{3 + 8 + 8 + 7 + 3 + 8} = \frac{263}{37} \approx 7.1$$

$$R_2 = \frac{3(-5) + 8(9) + 8(9) + 7(2) + 3(2) + 8(8)}{3 + 8 + 8 + 7 + 3 + 8} = \frac{213}{37} \approx 5.8$$

According to the *Little Mathematical Decision Formula*, then, with our choice of metrics, weights, and evaluations, the better restaurant overall for us is *Restaurant 1*, the livelier and less expensive restaurant closer to home. With a score of 7.1, it has been found overall to be very good, whereas *Restaurant 2*, with a score of 5.8, has been found only to be fairly good. According to this model, therefore, *Restaurant 1* should be chosen for our birthday dinner.

In closing, one must never forget that although the *LMDF* is powerful, it is not magical. Any decision produced by it is only as good as the steps that went before it. If the steps were *not* carried out properly, then the scores produced by the *LMDF* will have little or no connection to you personally and therefore will be largely meaningless, and any decision based on them will likely be unsound.

If, on the other hand, the steps were carried out correctly and accurately, then then final scores will represent true and accurate valuations of the options in accordance with your desires, tastes, judgments, values, and priorities, and the resulting decision will be eminently sound. Now, the scores won't be *perfect* nor the decision *infallible*—no such decision model exists because human judgments, opinions, and feelings can't be quantified with perfect precision. Nonetheless, the decision can and should be taken seriously in the decision process, just like sound advice from a wise and trusted friend, but it is advice that should be corroborated by common sense and intuition, as well as perhaps by wise and trusted friends of the flesh-and-blood variety.

Synopsis

Below is a brief synopsis of the decision procedure just described:

> **Step 1: Choose Metrics.** Consider your options carefully and identify all the criteria associated with them that have any relevance or importance to you for the decision at hand. Try to minimize their number by eliminating any redundant or superfluous ones, and

try to choose them as independent from one another as possible. Beware of decisions for which relevant information could be missing, as this can lead to missing metrics and unsound decisions.

Step 2: Assign Weights to Metrics. To each metric assign a number from 0 to 10 that represents as truly and accurately as possible its level of importance to you for the decision at hand. Beware of decisions involving metrics of extraordinary importance, since correct weights could then be difficult or impossible to assign, and unsound decisions would almost certainly result.

Step 3: Evaluate Options with Respect to Metrics. Evaluate each option with respect to each metric by choosing a number from -10 to +10 that represents as truly and accurately as possible the quality of that option for that metric in your judgment—disregarding the level of importance of the metrics in doing so. Beware of decisions involving options of unknown, uncertain, or extraordinary quality, as correct evaluations could then be difficult or impossible to perform, and unsound decisions would almost certainly result.

Step 4: Compute Scores & Identify Best Option. Finally, plug the weights and evaluations obtained in the previous two steps into the suitable version of the *Little Mathematical Decision Formula* for each option, and carefully compute their scores. These scores can then be utilized to identify the best option, to short-list options, to rank options from best to worst, or for whatever purpose one sees fit.

If these steps are carried out properly—and if the decision is suitable for the model in the first place—then the resulting decision will be eminently sound.

In the next chapter we take a look at how the *LMDF* can be used to make some typical decisions in life.

4

A Few Life Decisions

The *Little Mathematical Decision Formula* can be employed for many decisions in life, whether it be for identifying the best alternative, short-listing alternatives, or just ranking them from best to worst. It can be used to decide on a phone, computer, or tablet; a car, motorcycle, or bicycle; a suit, dress, or shoes. It can be used to decide on a job, college, or graduate school; a house, apartment, or condo; a restaurant, bar, or club. It can be used to decide on a plumber, carpenter, or electrician; a lawyer, accountant, or doctor; a mayor, governor, or president. Or it can be used to decide on a stock, bond, or mutual fund; a friend, boyfriend, or girlfriend; even a husband or a wife. To name just a few!

In this chapter we take a look at a small sampling of such decisions, and how the *LMDF* can be applied via the procedure described in the previous chapter. We look at six typical life decisions:

Deciding on a Car	Deciding on a House
Deciding on a Job	Deciding on a Stock
Deciding on a College	Deciding on a Significant Other

Each of these decisions is examined in turn in the sections below.

Deciding on a Car

One decision that can be challenging is deciding on a car. Obviously it is a fairly important decision representing a substantial personal and financial investment, but it can also be a complex decision requiring the consideration of a number of factors. The *Little Mathematical Decision Formula* can help, whether for deciding on a car to purchase, short-listing cars to be test-driven, or just ranking cars from best to worst.

Here's a typical example—one of many possible scenarios with this decision. Suppose we're trying to decide between two cars for purchase, *Car 1* which is a solid, reliable, economical car, and *Car 2* which is a fun, fast sports car.

We proceed with the four steps of the decision procedure. First we choose metrics, being careful to identify all the relevant ones, to choose the minimum number possible, and to choose those that are as independent from one another as possible:

Price	*Reliability*
Financing	*Size*
Appearance	*Gas Mileage*
Power	*Thrill Factor*
Handling	*Liking for Car*

Other metrics might have been chosen as well, such as a car's warranty, seating capacity, storage capacity, safety, comfort, prestige, delivery time, and so forth, but the metrics above are the ones we've deemed relevant for this decision.

Naturally, those with differing objectives, priorities, and experiences, as well as differing circumstances surrounding the decision, might choose very different metrics.

Those concerned largely with the economics of their purchase, for example, might simply choose the metrics of price, financing, warranty, reliability, and gas mileage, but ignore the others. Someone with children might be concerned, among other things, with seating capacity, storage capacity, safety, and comfort. And those purchasing a car outright obviously wouldn't care about financing. Whatever the case, the important thing is that one choose metrics that represent what is important *to them* for a decision.

Next we assign weights to the metrics, endeavoring to match them as closely as possible to the true level of importance we judge each metric to have:

Price	*8*	*Reliability*	*8*
Financing	*5*	*Size*	*5*
Appearance	*8*	*Gas Mileage*	*6*
Power	*9*	*Thrill Factor*	*8*
Handling	*9*	*Liking for Car*	*8*

From these numbers, it is evident that power and handling are extremely important to us in choosing a car, and that not far behind are price, appearance, reliability, thrill factor, and liking for a car. Gas mileage, financing, and size have only medium importance to us.

These weights would likely be quite different for someone with differing objectives, priorities, and experiences, as well as differing circumstances surrounding the decision. The largely economically-minded consumer, for example, would no doubt weigh more heavily than us the metrics of price, financing, reliability, and gas mileage, but would weigh less heavily, or perhaps omit altogether, the metrics of power, handling, appearance, thrill factor, and liking for a car. Those with children would likely weigh more heavily the metric of size, and they might include and weigh heavily additional metrics such as seating capacity, storage capacity, comfort, and safety; however, they might weigh less heavily the metrics of power, handling, appearance, and thrill factor. Whatever the case, the important thing is that one assign weights that reflect accurately the level of importance of each metric *to them*.

Next we evaluate each car with respect to the chosen metrics, endeavoring to match these evaluations as closely as possible to the true level of quality we judge each car to have:

	Car 1	*Car 2*
Price	*+9*	*-2*
Financing	*+7*	*+7*
Appearance	*+5*	*+9*
Power	*-5*	*+8*
Handling	*+3*	*+9*
Reliability	*+9*	*+7*
Size	*+8*	*+8*
Gas Mileage	*+9*	*0*
Thrill Factor	*-8*	*+8*
Liking for Car	*0*	*+7*

As these numbers indicate, *Car 1* for us is extremely good with respect to its price, reliability, and gas mileage; very good with respect to its financing and size; and fairly good with respect to its appearance and handling. On the other hand, it is fairly bad for its power, and very bad for its thrill factor. *Car 2* is for us extremely good with respect to its appearance and handling; and very good with respect to its power, size, reliability, financing, and thrill factor. It is neither good nor bad with respect to its gas mileage, and slightly negative for its price. Regarding our feeling for the cars, we neither like nor dislike *Car 1*, but we like *Car 2* quite a lot.

Naturally, these evaluations could be quite different for someone with different tastes and preferences, or different circumstances surrounding the decision. A wealthy individual, for example, might judge even *Car 2* to be relatively inexpensive and give it a much higher evaluation for price than we did, but someone not-so-wealthy might give it an even lower evaluation. A true sports-car enthusiast might judge the power and handling of even *Car 2* to be deficient and give it lower marks than us in these respects. And of course what is a good-looking car for one might be ugly for another; and a car that one person loves another might hate. Whatever the case, the important thing in the end is that one's evaluations capture accurately the level of quality of each option *to them.*

Finally, we plug the weights and evaluations into the suitable version of the *LMDF* for *Car 1* and *Car 2* and compute their scores:

$$C_1 \approx 3.0 \qquad\qquad C_2 \approx 6.2$$

According to the *Little Mathematical Decision Formula*—with our choice of metrics, weights, and evaluations—the better car for us overall is *Car 2*, the fun, fast sports car. With a score of 6.2, it has been found to be quite good overall, whereas *Car 1*, with a score of 3.0, has been found to be only decent. According to this model, therefore, *Car 2* should be purchased.

If the steps above have been carried out correctly and accurately, then these scores will represent true and accurate overall valuations of the cars according to our desires, tastes, judgments, values, and priorities, and therefore the resulting decision to purchase *Car 2* will be eminently sound, and can and should be taken seriously in the decision process. If the steps haven't been carried out properly, however, then the scores will be largely meaningless, and any decision based on them will be unreliable.

Deciding on a Job

Another often challenging decision in life is choosing a job or a career. Obviously it is an important decision having significant personal and financial ramifications, but it can also be a complex decision for which numerous issues must be considered. The *Little Mathematical Decision Formula* can help, whether for short-listing jobs to which to apply, or for deciding on a job to actually accept.

Here's a typical example—one of many possible scenarios with this decision. Suppose we're a young, single adult, and we're trying to decide between two jobs, *Job 1* being a high-paying, high-stress job with long hours, and *Job 2* a lower-paying, lower-stress job with shorter hours.

We proceed with the four steps of the decision procedure. First we choose metrics, being careful to identify all the relevant ones, to choose the minimum number possible, and to choose those that are as independent from one another as possible:

Salary	*Hours*
Benefits	*Stress*
Upward Mobility	*Commute*
Work Environment	*Vacation Time*

Other metrics might have been chosen as well, such as the difficulty of the work, the enjoyability of the work, the likeability of coworkers, the likeability of the boss, the prestige of the job, the perks, and so forth, but the metrics above are the ones we've deemed relevant for this decision.

As always, these metrics could vary significantly depending on one's objectives, priorities, and experiences, as well as the circumstances of the decision. Those desiring a job solely for the money, for example, might choose only the metrics of salary and benefits, and ignore the hours, enjoyability, and vacation time. Someone independently wealthy might be concerned only with quality-of-life factors such as hours, enjoyability, stress, work environment, and vacation time, but ignore salary, benefits, and upward mobility. Those desiring a job that accomplishes good in the world would no doubt include a metric for the altruistic character of the work. Whatever the case, the important thing is that one choose metrics that represent what is important *to them* for a decision.

Next we assign weights to the metrics, endeavoring to match them as closely as possible to the true level of importance we judge each metric to have:

Salary	*8*	*Hours*	*3*
Benefits	*5*	*Stress*	*8*
Upward Mobility	*6*	*Commute*	*7*
Work Environment	*5*	*Vacation Time*	*6*

From these numbers, it is apparent that salary and stress level are very important factors for us in choosing a job, and that not far behind is the commute. Still fairly important are upward mobility and vacation time, and only slightly less important are benefits and work environment. Hours, however, are of fairly low importance to us for a job.

As always, these weights would likely be quite different for someone with differing objectives, priorities, and experiences, or differing circumstances surrounding the decision. The largely money-minded, for example, would likely weigh even more heavily than us the metrics of salary and benefits, but perhaps weigh less heavily, or even omit altogether, the metrics of hours, stress, commute, and vacation time. Someone semi-retired or financially independent might weigh less heavily, or even omit altogether, the metrics of salary, benefits, and upward mobility, but weigh more heavily the quality-of-life metrics of hours, enjoyability, and stress. Those desiring a job that accomplishes good in the world might weigh less heavily the metrics of salary, benefits, upward mobility, hours, and stress, but would include and weigh heavily a metric for the altruistic character of the work. Whatever the case, the important thing is that one assign weights that capture accurately the level of importance of each metric *to them.*

Next we evaluate each job with respect to the chosen metrics, taking care to match the evaluations as closely as possible to the true level of quality we judge each job to have:

	Job 1	*Job 2*
Salary	+9	-3
Benefits	+7	-3
Upward Mobility	+7	+5
Work Environment	+5	+8
Hours	-7	+5
Stress	- 5	+8
Commute	-4	+5
Vacation Time	+4	+7

From these numbers, it is evident that for us *Job 1* is extremely good with respect to its salary, very good with respect to its benefits and upward mobility, and fairly good with respect to its work environment and vacation time. On the other hand, it is fairly bad for its stress and commute, and very bad for its hours. *Job 2* is for us very good with respect to its work environment, stress, and vacation time; and fairly good with respect to its upward mobility, hours, and commute. It is, however, fairly bad for its salary and benefits.

Naturally, these evaluations could be very different for someone with differing tastes and preferences, or differing circumstances surrounding the decision. Someone accustomed to a very large salary, for instance, might judge even the salary of *Job 1* to be relatively low, and that of *Job 2* to be dismal, and might therefore give them much lower marks than we did. A hard worker accustomed to long hours might judge even *Job 1* to be fine in this respect and give it a higher evaluation, but a less motivated worker might give it an even lower rating than us. And of course what constitutes a good work environment for one might be intolerable for another, and what is a stressful job for one might be exciting and challenging for another. Whatever the case, the important thing is that one's evaluations reflect accurately the level of quality of each option *to them*.

Finally, we plug the weights and evaluations into the suitable version of the *LMDF* for *Job 1* and *Job 2* and compute their scores:

$$J_1 \approx 2.3 \qquad\qquad J_2 \approx 3.9$$

According to the *Little Mathematical Decision Formula*—with our choice of metrics, weights, and evaluations—the better job for us overall is *Job 2*, the lower-paying, lower-stress job with shorter hours. With a score of 3.9, it has been found to be fairly good overall, whereas *Job 1*, with a score of 2.3, has been found to be only slightly good.

Regarding a decision, though *Job 2* is clearly the better choice according to this model, we might judge that its overall quality is too low. If this be the case, then naturally we should consider other jobs and try to find a better fit. If, on the other hand, we judge that this job is still worth pursuing—or if we've no choice in the matter—then obviously we should pursue it.

In any event, if the steps above have been carried out correctly and accurately, then the scores obtained will represent true and accurate overall valuations of the jobs according to our desires, tastes, judgments, values, and priorities; therefore, either decision above will be eminently sound, and can and should be taken seriously in the decision process. If the steps haven't been carried out properly, however, then the scores will be largely meaningless, and any decision based on them will be unreliable.

Deciding on a College

Choosing a college or university is another often daunting decision. It is obviously an important one having lifelong consequences, but it can also be quite complicated, requiring careful consideration of many factors. The *Little Mathematical Decision Formula* can help, whether for short-listing schools to be applied to, or making the final decision of which to attend.

Here's a typical example. Suppose we're trying to decide between two universities, having already been accepted to both. *University 1* is very large and highly ranked, and is located in a major city not too far from home. *University 2* is not as large but is even more highly ranked, and is located in a relatively small college town quite far from home.

We proceed with the usual four steps. First we choose metrics, endeavoring to identify all the relevant ones, to choose the minimum number possible, and to choose those that are as independent from one another as possible:

Tuition	*Campus Life*
Financial Aid	*Location*
Ranking	*Climate*
Size	*Proximity to Home*
Academics	*Beauty of Campus*
Athletics	*Liking of School*

Though these are a fairly large number of metrics, we were unable to reduce their number without omitting something important—so we embrace the complexity of this decision! Of course, other metrics might have been chosen we well, such as class size, student-to-professor ratio, living options, dining options, stature of a particular department, fraternity or sorority options, work opportunities, and so forth. The metrics above, however, are the ones we consider important for this decision. In *Chapter 5*, incidentally, a simplification of the present model is discussed for which only the most important metrics are included.

As always, those with differing objectives, priorities, and experiences, as well as differing circumstances surrounding the decision, might choose very different metrics. A wealthy student, for example, might omit altogether the metrics of financial aid and tuition, whereas a not-so-wealthy student would almost certainly include these, plus perhaps a metric for work opportunities. Many students would include a metric for fraternity or sorority options. A student already focused on a particular field of study might include a metric for the stature of a particular department. More practical-minded students might omit altogether the subjective metrics of beauty of campus and liking of a school. And of course not every student cares about proximity to home, climate, or rankings. Whatever the case, the most important thing is that one choose metrics that represent what is important to *them* for a decision.

Next we assign weights to the metrics, endeavoring to match them as closely as possible to the true level of importance we judge each metric to have:

Tuition	*8*	*Campus Life*	7
Financial Aid	*8*	*Location*	5
Ranking	*5*	*Climate*	7
Size	*5*	*Proximity to Home*	3
Academics	*9*	*Beauty of Campus*	7
Athletics	*3*	*Liking of School*	7

From these numbers, it is evident that for us academics is extremely important in choosing a college; tuition and financial aid are very important; and not far behind are campus life, climate, beauty of campus, and liking of school. Of only medium importance to us are ranking, size, and location; and of relatively low importance are athletics and proximity to home.

These weights, of course, would likely be quite different for someone with differing objectives, priorities, and experiences, as well as differing circumstances surrounding the decision. A wealthy student, for example, might place much less weight on financial aid and tuition, but a not-so-wealthy student might place even more weight on these metrics than we did. An avid hiker, skier, surfer, or golfer might place much more weight than us on location and climate. Student athletes would almost certainly weigh athletics much more heavily than we did. And largely practical-minded students might weigh much less heavily, or perhaps omit altogether, the metrics of beauty of campus and liking of a school. In any case, what really matters in the end is that one assign weights that capture accurately the level of importance of each metric *to them*.

Next we evaluate each college with respect to the chosen metrics, endeavoring to match the evaluations as closely as possible to the true level of quality we judge each school to have:

	University 1	*University 2*
Tuition	-3	-6
Financial Aid	+7	+7
Ranking	+6	+9
Size	-3	+8
Academics	+8	+8
Athletics	+8	+8
Campus Life	+7	+7
Location	+8	+5
Climate	-2	+7
Proximity to Home	+5	-5
Beauty of Campus	-3	+8
Liking of School	+3	+8

As these numbers indicate, *University 1* for us is very good with respect to its academics, athletics, and location; nearly as good for its financial aid and campus life; and fairly good for its ranking and proximity to home. Its tuition, size, beauty, and climate, on the other hand, are somewhat negative. *University 2* is extremely good for us with respect to its ranking; very good for its size, academics, athletics, and beauty; nearly as good for its financial aid, campus life, and climate; and fairly good with respect to its location. Its tuition and proximity to home, however, are quite negative for us. Regarding our feeling about the schools, we like *University 1* somewhat, but *University 2* we like very much.

These evaluations, naturally, could be very different for someone with differing tastes and preferences, or differing circumstances surrounding the decision. For us, being a very large school is a negative, but for others it will be a positive. In the same vein, being located in a major city is highly desirable for us, but for another it will be undesirable. And what constitutes affordable tuition for one will be unaffordable for another; what constitutes exciting and stimulating campus life for one will be of no interest to another; and what is close to home for one will be faraway for another. Whatever the case, the most important thing in the end is that one's evaluations accurately capture the level of quality of each option *to them.*

Finally, we plug the weights and evaluations into the suitable version of the *LMDF* for *University 1* and *University 2* and compute their scores:

$$U_1 \approx 3.0 \qquad\qquad U_2 \approx 5.5$$

According to the *Little Mathematical Decision Formula*—with our choice of metrics, weights, and evaluations—the school that is the better fit for us overall is *University 2*, the smaller, more highly ranked school in a small college town far from home. With a score of 5.5, it has been found to be quite good overall, whereas *University 1*, with a score of 3.0, has been found only to be fairly good.

Regarding a decision, we're in the same situation here as with the previous example. Indeed, though *University 2* is clearly the better school according to this model, we might judge that its overall quality is too low. If this be the case, then naturally we should, if possible, consider other colleges and attempt to find one that's a better fit. If, on the other hand, we judge that this school is still worth attending—or if we've no choice in the matter—then obviously we should enroll.

In any event, if the steps above have been carried out correctly and accurately, then the scores obtained will represent true and accurate overall valuations of the colleges according to our desires, tastes, judgments, values, and priorities; therefore, either decision above will be eminently sound, and can and should be taken seriously in the decision process. If the steps haven't been carried out properly, however, then the scores will be largely meaningless, and any decision based on them will be unreliable.

Deciding on a House

A very important decision that can be difficult and complex is the purchase of a house. The *Little Mathematical Decision Formula* can help, whether for short-listing houses to visit, or making the final decision of which to purchase.

Here's a typical example—one of many possible scenarios with this decision. Suppose we're young, married, with a good job and no children, and that we're looking to purchase our first home. Suppose we're trying to decide between two houses, *House 1* which is a relatively new, fairly small house close to the center of a medium-sized town, and *House 2* which is an older, larger house on the outskirts of town.

We proceed with the usual four steps. First we choose metrics, being careful to identify all the relevant ones, to choose the minimum number possible, and to choose those that are as independent from one another as possible:

Price	*Room Layout*
Financing	*Property*
Taxes	*Neighborhood*
Size	*Surrounding Town*
Quality	*Proximity to Retail*
Attractiveness	*Liking for House*

Though these are a fairly large number of metrics, we were unable, as in the previous example, to reduce their number without omitting something important, so we again accept the complexity of this decision and go with all of them. Naturally, other metrics might have been chosen as well, such as age, style, coziness, quality of schools, proximity to work, proximity to family or friends, resale potential, and so forth, but the metrics above are the ones we consider important for this decision. In *Chapter 5* a simplification of the present model is discussed for which only the most important metrics are included.

These metrics could, as always, vary significantly depending on one's objectives, priorities, and experiences, as well as the circumstances of the decision.

Those with children, for example, might include metrics for the quality of schools, proximity to schools, safety of the neighborhood, and perhaps the number of children already living there. A retired individual might include metrics such as proximity to their children and grandchildren, proximity to health facilities, or how many levels a house entails. Someone wealthy buying a house outright would obviously not be concerned with financing, and might be unconcerned with price and taxes as well. And those purchasing a house strictly for investment—to fix and flip it—would likely be greatly interested in resale potential. In any case, what really matters is that one choose metrics that represent what is important *to them* for a decision.

Next we assign weights to the metrics, endeavoring to match them as closely as possible to the true level of importance we judge each metric to have:

Price	*8*	*Room Layout*	*8*
Financing	*8*	*Property*	*5*
Taxes	*7*	*Neighborhood*	*9*
Size	*7*	*Town*	*8*
Quality	*7*	*Proximity to Retail*	*5*
Attractiveness	*5*	*Liking for House*	*7*

From these numbers, it is evident that the neighborhood surrounding a house is extremely important to us; price, financing, room layout, and the surrounding town are very important; and not far behind are taxes, size, quality, and our liking for a house. Of only medium importance to us are a house's attractiveness, property, and proximity to retail.

These weights, as always, could be quite different for someone with differing objectives, priorities, and experiences, as well as differing circumstances surrounding the decision. Someone wealthy, for example, might weigh much less heavily the metrics of price, financing, and taxes, and might even omit one or more of them entirely; but someone not-so-wealthy might weigh them even more heavily than we did. Those purchasing a house strictly to fix and flip it would likely be even more concerned than us about price and quality, and would almost certainly include and weigh heavily an additional metric for resale potential. A professional architect or builder might place much more weight on quality, attractiveness, and room layout than we did. Whatever the case, the important thing is that one assign weights that capture accurately how important each metric is *to them.*

Next we evaluate each house with respect to the chosen metrics, endeavoring to match the evaluations as closely as possible to the true level of quality we judge each house to have:

	House 1	*House 2*
Price	*-3*	*+5*
Financing	*+8*	*+8*
Taxes	*-5*	*+5*
Size	*+8*	*+3*
Quality	*+9*	*+5*
Attractiveness	*+8*	*+5*
Room Layout	*+7*	*+5*
Property	*+3*	*+7*
Neighborhood	*+7*	*+7*
Town	*+8*	*+8*
Proximity to Retail	*+8*	*+3*
Liking	*+8*	*+5*

From these numbers, it is apparent that for us *House 1* is extremely good with respect to its quality; very good with respect to its financing, size, attractiveness, surrounding town, and proximity to retail; nearly as good for its room layout and neighborhood; and only decent with respect to its property. Its price, on the other hand, is somewhat of a negative for us, and its taxes are even more so. As for

House 2, it is very good for us with respect to its financing and surrounding town; nearly as good for its property and neighborhood; fairly good for it price, taxes, quality, attractiveness, and room layout; and only decent for its size and proximity to retail. Regarding our feeling for the houses, we like *House 2* to some degree, but *House 1* we like very much.

As always, these evaluations could vary significantly depending on one's tastes and preferences, as well as the circumstances of the decision. Someone wealthy, for example, might judge both houses to be quite inexpensive to purchase and carry, and might give them much higher marks for price and taxes; but a not-so-wealthy person might give them even lower marks than we did. A professional architect, designer, or builder having very high standards would likely give the houses lower ratings for quality and attractiveness than we did. Those accustomed to living in a major city might find the surrounding area of both houses to be fairly boring and thus give them a lower rating than us with respect to this metric. And of course what is a spacious house for one will be cramped for another; what is a large piece of property for one will be small for another; and what is a good neighborhood for one will be not-so-good for another. Whatever the case, the important thing in the end is that one's evaluations accurately reflect the level of quality of each option *to them*.

Finally, we plug the weights and evaluations into the suitable version of the *LMDF* for *House 1* and *House 2* and compute their scores:

$$H_1 \approx 5.45 \qquad\qquad H_2 \approx 5.62$$

According to the *Little Mathematical Decision Formula*—with our choice of metrics, weights, and evaluations—the better house for us overall by a tiny margin is *House 2*, the older, larger house on the outskirts of town. With a score of 5.62, it has been found to be fairly good overall, and *House 1*, with a score of 5.45, has been found to be of only slightly lower quality.

Regarding a decision, we must for all practical purposes consider the houses to be *tied*. The reason, as was discussed in *Step 4* of the previous chapter, is the small imprecision always present when determining weights and evaluations, due to the fact that it is simply not possible to assign perfectly precise numerical values to human judgments, opinions, and feelings. Indeed, the scores for *House 1* and *House 2* are so close in value that even a very small difference in the weights and evaluations could produce enough of a change in the final scores to change the decision above. In fact, mathematically, merely a *tenth of a point* difference in these quantities could cause the decision to change. No one has such a precise bead on such judgments! In the final analysis, then, though this model has produced mathematically precise scores and an apparently definitive decision, the small human error inherent in *applying* the model makes this decision in reality too close to call, and hence effectively a tie. It *is* a tie, in fact, in a very real sense, because given that identifying a truly better house would require precision in the weights and evaluations beyond human ability, the houses are in fact indistinguishable in our eyes and hence of essentially equal overall quality. In the end, then, we wouldn't go wrong with either choice! Unless of course we judge the scores to be too low, in which case we should consider other houses and attempt to find a better one.

Deciding on a Stock

Choosing a stock is an interesting and important decision that with some care and effort can be addressed using the *Little Mathematical Decision Formula*.

Here's a typical example. Suppose we're looking for a stock for the purpose of investment, that is, a stock that likely will be safe to hold for a period of time and yield a solid return over that time. We are not, in other words, looking for a speculative stock to buy and sell quickly in order to make a sizeable profit. Suppose, then, we've identified two stocks, *Stock 1* which has been doing extremely well lately, and *Stock 2* which hasn't done particularly well recently, but has been solid and steady for years. We'll decide between these two stocks.

We proceed with the usual four steps, beginning by choosing metrics. Now, not being expert at stock selection, we appeal to five traditional metrics for picking stocks of the desired type. These metrics concern certain fundamental characteristics of a company represented by a stock and their relation to the stock's price:

Size
Financial State
Earnings History
Dividends History
Stock Valuation

Size refers to a company's total assets, and its *financial state* refers to the ratio of its assets to its liabilities. *Earnings history* and *dividends history* refer to the amount and growth of each annually as demonstrated by a company during its existence. *Stock valuation* refers to a company's price-to-earnings and price-to-assets ratios. These quantities will together determine how good or bad a stock is with respect to the present decision.

Other metrics are also possible for this kind of decision, such as the quality of management, the identity of the largest stockholders, the health of the sector represented by a stock, the health of the stock market generally, the health of the larger economy, and so on. Indeed, for a decision as complex as stock selection, the potentially relevant factors are virtually limitless! Given our lack of expertise, however, we happily limit ourselves to the classic five metrics above.

Naturally, investors with different objectives and priorities, as well as different levels of expertise, would likely choose different metrics. Those adopting a more speculative approach, for example, would almost certainly choose them differently, as would those seeking to create a more diversified portfolio. A long-time expert in equities would almost certainly choose differently—and different experts will naturally have different opinions on the matter! Whatever the case, the important thing, as always, is that one uses one's best judgment to select the metrics that represent what is truly important for a decision.

Next we assign weights to the metrics. In actual practice, this would require one to analyze the financial accountings represented by the metrics and ascertain their respective levels of importance with regards to the present decision. To keep things simple, however, for the sake of illustration, we simply assign a weight of eight to each metric:

Size	*8*
Financial State	*8*
Earnings History	*8*
Dividends History	*8*
Stock Valuation	*8*

These weights obviously impart equally high importance to each metric, a perhaps not entirely unreasonable assumption for this decision as a first approximation. Notwithstanding, we encourage any reader serious about such a decision to do the requisite research or consultation and come up with a more refined set of weights.

Next we evaluate each stock with respect to the chosen metrics. In actual practice, doing so would require one to acquire the necessary information about each company as dictated by the metrics, and then to ascertain the various quality levels of the stocks as indicated by this information. For instance, regarding the metric of *Stock Valuation*, one would have to obtain the price-to-earnings and price-to-assets ratios for each company, and then ascertain how good or bad the corresponding stocks were with respect to this metric for the present decision. If, for example, a company's price-to-earnings and price-to-assets ratios over the past couple years were greater than 30 and 3, respectively, then its stock would traditionally be considered quite overvalued and therefore of quite negative quality with respect to this metric for this decision. To keep things simple, however, for the sake of illustration, we suppose that all this has been done, and that the following evaluations have been determined:

	Stock 1	*Stock 2*
Size	-5	+7
Financial State	0	+5
Earnings History	-2	+8
Dividends History	-5	+8
Stock Valuation	-4	+6

From these numbers it is readily apparent that the company represented by *Stock 1* has been found seriously lacking in every department. Its financial state is its best quality, and this is merely neutral, neither good nor bad. All its other characteristics are negative, with its earnings history being only slightly so, and its stock valuation, dividend history, and size being substantially so. The company represented by *Stock 2*, on the other hand, has been found to be of good quality across the board. Its financial state and stock valuation are fairly good, its size is very good, and its earnings and dividends history are excellent.

Different investors, naturally, would likely evaluate the stocks differently. Those adopting a less conservative approach than us, for example, would likely assign higher marks to both stocks, and in their eyes even *Stock 1* might not seem so bad. On the other hand, investors espousing an even more conservative approach would likely assign lower marks, and might even be skeptical of our highly-rated *Stock 2*. Of course, if one's basic financial objective is different—say if one is looking to speculate for a quick profit—then one's evaluations would likely be quite different, to say nothing of the metrics themselves.

Finally, the scores for *Stock 1* and *Stock 2* can be computed. Plugging the weights and evaluations into the suitable version of the *LMDF* for each stock, then, we obtain:

$$S_1 = -3.2 \qquad\qquad S_2 = 6.8$$

According to the *Little Mathematical Decision Formula*—with our choice of metrics, weights, and evaluations—the better stock for us overall by a large margin is *Stock 2*. With a score of 6.8, it has been found to be quite good overall, whereas *Stock 1*, with a score of – *3.2*, has been found to be quite poor. According to this model, therefore, *Stock 2* should be purchased, unless we judge that its overall quality is too low, in which case we should consider other stocks and attempt to find a better one.

In any event, if the steps above have been carried out correctly and accurately, then the scores will represent true and accurate overall valuations of the stocks based on facts and expert judgment, and therefore either decision will be eminently sound, and can and should be taken seriously in the decision process. If, however, the steps have not been carried out properly, then the scores will be largely meaningless, and any decision based on them will be unreliable.

Deciding on a Significant Other

Finally, the *Little Mathematical Decision Formula* can help even in deciding on a girlfriend, boyfriend, husband, wife, partner, or a significant other of any kind—an obviously important and often complicated decision! Here's a typical example. Suppose we're single, never married, with no children, and that we're trying to choose between two individuals—*Suitor 1* and *Suitor 2*—for a serious long term relationship.

We proceed with the usual four steps. First we choose metrics, being careful to identify all the relevant ones, to choose the minimum number possible, and to choose those that are as independent from one another as possible:

Age	*Honesty*
Beauty	*Authenticity*
Friendliness	*Reliability*
Sexiness	*Faithfulness*
Intelligence	*Common Interests*
Kindness	*Liking*
Communicativeness	*Love*

These are a lot of metrics! Try as we may, however, we are unable to reduce their number without omitting something important—so once again we embrace the complexity of this decision! Of course, many other metrics might have been chosen as well, such as sense of humor, fun-lovingness, warmth, generosity, humility, loyalty, cooperativeness, talent, education, profession, social status, financial status, and so on. The metrics above, however, are the ones we've deemed relevant for this decision. In *Chapter 5*, a simplification of the present model is discussed for which only the most important metrics are included.

As always, the metrics chosen depend completely on who has chosen them—one's objectives, priorities, and experiences, as well as the circumstances of the decision. Some, for example, would include the metrics of sense of humor and fun-lovingness, and others would include metrics for the financial and social status of their potential partner. Someone having been in a toxic relationship might include metrics relating to anger, jealousy, or possessiveness. And someone seeking a husband or wife would likely include metrics such as stability, loyalty, or maternal or paternal qualities. Whatever the case, the important thing in the end is that one choose metrics that represent what is important *to them* for a decision.

Next we assign weights to the metrics, being sure they match as closely as possible the true level of importance we judge each metric to have:

Age	*5*	*Honesty*	*10*
Beauty	*7*	*Authenticity*	*10*
Friendliness	*7*	*Reliability*	*9*
Sexiness	*8*	*Faithfulness*	*10*
Intelligence	*7*	*Common Interests*	*8*
Kindness	*8*	*Liking*	*8*
Communicativeness	*9*	*Love*	*9*

From these numbers, it is evident that for us honesty, authenticity, and faithfulness are extremely important for a serious relationship, and that nearly as important are communicativeness, reliability, and love. Still very important to us are sexiness, kindness, common interests, and our liking of the person; and not far behind are beauty, friendliness, and intelligence. Age has only medium importance to us for a potential partner.

These weights, as always, could be quite different for different individuals depending on their objectives, priorities, and experiences, or the circumstances of the decision. Those, for example, whose first and foremost concern is beauty would likely weigh this metric even more heavily than we did, and they might weigh less heavily metrics pertaining to personality or character. Someone for whom social and financial status is all-important would no doubt include and weigh heavily these metrics, and they might weigh less heavily or perhaps omit altogether metrics pertaining to appearance or personality. Those seeking a casual relationship might weigh less heavily than us the metrics of reliability, faithfulness, and love, but they might weigh even more heavily beauty and sexiness. Finally, those to whom age is unimportant would weigh this metric even more lightly than us, but those to whom it is very important would weigh it much more heavily. Whatever the case, the important thing in the end is that one assign weights that accurately reflect how important each metric is *to them.*

Next we evaluate each suitor with respect to the chosen metrics, endeavoring to match the evaluations as closely as possible to the true level of quality we judge each person to have:

	Suitor 1	*Suitor 2*
Age	*+9*	*-2*
Beauty	*+9*	*+6*
Friendliness	*+9*	*+8*
Sexiness	*+9*	*+5*
Intelligence	*+7*	*+8*
Kindness	*+6*	*+7*
Communicativeness	*+8*	*+4*
Honesty	*+8*	*+9*
Authenticity	*+8*	*+9*
Reliability	*+8*	*+9*
Faithfulness	*+7*	*+8*
Common Interests	*+8*	*+7*
Liking	*+7*	*+7*
Love	*+8*	*+7*

From these numbers, it is apparent that for us *Suitor 1* is extremely good with respect to age, beauty, friendliness, and sexiness; very good for communicativeness, honesty, authenticity, reliability, and common interests; and nearly as good for intelligence and faithfulness. *Suitor 2* is for us extremely good with respect to honesty, authenticity, and reliability; very good with respect to friendliness, intelligence, and faithfulness; nearly as good for kindness and common interests; and fairly good for beauty, sexiness, and communicativeness. The age of *Suitor 2*, however, is a slight negative for us. Regarding our feelings for these potential partners, we like them both very much, and we love them very much as well, perhaps *Suitor 1* a little more.

As always, these evaluations could be very different for those with differing tastes and preferences, as well as differing circumstances surrounding the decision. Someone of a very different age than us, for example, might evaluate the suitors very differently in this respect, giving our ideally-aged *Suitor 1* much lower marks, and our not-so-ideally aged *Suitor 2* much higher ones. A brilliant person accustomed to being around other brilliant people might bestow upon even our very intelligent candidates lower ratings than us in this regard. A quiet, insular individual might find the friendliness and communicativeness of our suitors to be intrusive and give them lower marks with respect to these metrics. And of course beauty, sexiness, and love are totally in the eyes and heart of the beholder. Whatever the case, the important thing in the end is that one's evaluations accurately reflect the level of quality of each option *to them*.

Finally, we plug the weights and evaluations into the suitable version of the *LMDF* for *Suitor 1* and *Suitor 2* and compute their scores:

$$S_1 \approx 7.9 \qquad\qquad S_2 \approx 6.9$$

According to the *Little Mathematical Decision Formula*—with our choice of metrics, weights, and evaluations—the better prospective partner for us overall by a fairly small margin is *Suitor 1*. With a score of 7.9, he or she has been found to be very good overall, whereas *Suitor 2*, with a score of 6.9, though still very good, is not quite as good as *Suitor 1*.

Regarding a decision, though the one-point margin between these scores is on the borderline, as long as the steps have been carried out correctly and accurately, the scores will represent sufficiently true and accurate overall valuations of the suitors, and the decision to choose *Suitor 1* will be sound. If, however, the steps haven't been carried out properly, then the indicated decision will not be reliable.

5

Alternative Decision Models

In this final chapter we take a brief look at a few alternative models for decision-making based on the *Little Mathematical Decision Formula*. Besides offering the reader additional models to consider and perhaps experiment with, learning about these models will enhance one's understanding of the present model and of decision-making in general, and will provide a deeper glimpse into the interesting and important process of mathematical modeling. We consider five such models, respectively, in the sections below:

1. Un-Weighted Average
2. Most Important Metrics Only
3. Alternate Scaling
4. Decisions with Uncertainty
5. Decisions with Greater Uncertainty

These models are essentially *variants* of the present model and represent various efforts to improve it, either by simplifying it or by broadening its applicability. Tradeoffs, however, almost always exist in mathematical modeling. Simpler models tend to be more limited in their applicability, and more broadly applicable ones tend to be more complicated and difficult to implement. In the final analysis, then, though these variants have their place, the present model represents an excellent balance between simplicity and applicability, a sweet spot as it were that is hard to improve upon.

Un-Weighted Average

The *Little Mathematical Decision Formula*, as we're well aware by now, represents a sophisticated kind of averaging process called a weighted average. One way to simplify this model, then, is to replace the weighted average with a simple *un-weighted* average—in effect making it a weighted average whose weights are all equal to one.

The resulting decision model is almost identical to the original, only lacking a step for choosing weights. There are thus three steps. First, one chooses the appropriate metrics exactly as before. Next, one evaluates each option with respect to these metrics, as before. Finally, one computes a simple average of these evaluations for each option, from which a decision can then be made.

The upside of this alternate model is that it is substantially simpler than the original, eliminating the necessity of choosing weights and permitting a simpler computation to boot. Its considerable downside, however, as we saw in *Chapter 2*, is that an un-weighted average represents an extreme over-simplification of the decision-making process and will produce incorrect scores and unsound decisions in almost every situation. In short, this alternative model represents a dismal tradeoff—somewhat greater simplicity for an extreme reduction in applicability—and therefore it is not at all preferable to our original model.

Most Important Metrics Only

The *Little Mathematical Decision Formula* as implemented in this book requires that every metric with any relevance at all be included. As we've seen, however, this can sometimes result in a large number of metrics and correspondingly long formulas and computations. An easy way to simplify this model, then, is to incorporate fewer metrics, in particular only the *most important metrics*, and to ignore the rest.

The resulting decision model is almost identical to the original, requiring only a slight modification of the first step. We briefly describe the steps. First, one chooses metrics—but now only the *most important* ones. This can be done in a few different ways. First, one can simply choose the metrics that one judges to have the greatest importance for a decision. If, however, one wants to be a bit more scientific, one can first assign weights to all the possible metrics and then choose those that have the largest weights. To this end, one can either choose the metrics that have weights larger than some predetermined value—say 7 or 8—or those that fall above some natural division in the weights. In any case, once the metrics are chosen, then one proceeds exactly as before, by assigning weights to the metrics, evaluating the options with respect to these metrics, and finally computing the scores for each option using the *LMDF*—from which a decision can then be made.

The upside of this alternative model is that it is simpler than the original, not requiring one to identify every metric that has any relevance at all, and also permitting a simpler computation due to the smaller number of metrics. Its downside, however, is that if a decision is close—if the scores produced by the *LMDF* using *all* the relevant metrics would be close in value—then omitting even the least weighted metrics could distort these scores enough to incorrectly change the decision. In the final analysis, then, we prefer our original model, because the greater simplicity of this variant doesn't seem to be worth the price of the incorrect decisions produced when decisions are close.

Alternative Scaling

The original decision model employs numerical scales from 0 to 10 for metrics and from -10 to +10 for evaluations, and it includes specific interpretations of these numerical values. Now, as we know, these scales and interpretations work very well for many decisions in life, but not for all of them; in particular, they don't work for decisions involving weights representing off-the-scale importance or evaluations representing off-the-scale quality. For such decisions, then, the original model doesn't apply.

The simplest way to remedy this and create a more broadly applicable model is to devise a different scaling for weights and evaluations that can accommodate such off-the-scale values. For example, a scale from 0 to 1000 could be used for weights, and from -1000 to +1000 for evaluations. With these scales, ordinary weights and evaluations could still be represented by small numbers as before, but those representing extraordinary, off-the-scale importance or quality could now be accommodated by the larger numbers.

The resulting decision model is almost identical to the original, and consists of essentially the same four steps. First, one chooses the appropriate metrics as before. Next, one assigns the suitable weights to each metric—but now on a scale from 0 to 1000. Next, one evaluates each option with respect to these metrics—but now on a scale from -1000 to +1000. Finally, one computes the scores for each option using the *LMDF* exactly as before, from which a decision can then be made.

The good news about this alternative model is that it applies to a greater number of decisions than the original. The bad news, however, is that its modified scaling turns out to be substantially more difficult to implement in practice—and exactly for those problematic decisions it was designed to accommodate!

A simple example makes this clear. Suppose we're attempting to assign a numerical weight to a metric of extremely critical importance, such as the competence of a physician, or the safety of a ship or a plane. Should this weight be chosen to be 100? Or 500? Or 1000? Or should we devise an even broader scale to accommodate still larger numbers? The answer is not at all clear. Generally, attempting to assign the proper numerical weight to a metric of extraordinary importance can be very difficult even with these re-scalings—and a dangerous game when making extremely critical decisions.

In the final analysis, then, though this alternate model has broader applicability *in theory*, in actual practice it doesn't work out that way at all. We therefore prefer our original, easier-to-implement model, and leave those life-or-death decisions to other means.

There is, incidentally, another way of remedying the scaling problem of our original model, namely, by retaining the original numerical scales but changing the *interpretation* of the numbers on that scale. Ordinary weights and evaluations, for instance, can be represented by small numbers between 0 and ±1, and those representing extraordinary importance or quality can be represented by numbers closer to ±10. Mathematically, this is equivalent to the variant described above, but since it is equally difficult to implement, and not even as intuitive to boot, we consider it even less preferable.

Decisions with Uncertainty

The *Little Mathematical Decision Formula*, as we're well aware, is not equipped to handle decisions involving uncertainty, namely, decisions involving alternatives that have uncertain quality with respect to any metric. A simple example illustrates the problem. Suppose we're deciding between jobs and are employing the metric of *salary*, and suppose moreover that the salary for one of the jobs is *uncertain*—say either $35,000 or $50,000. How, then, are we to evaluate this job with respect to its salary? Unless we feel equally about these salaries—most obviously wouldn't—then there simply cannot be a single numerical evaluation to be plugged into the *LMDF*. This model, therefore, cannot be applied to this decision, and it is likewise inapplicable to any decision involving the same kind of uncertainty.

One possible remedy would be to evaluate the two salaries separately and then simply insert their *average* into the *LMDF*. For example, if we judged the salary of $35,000 to be decent and assigned it an evaluation of +5, and we judged the salary of $50,000 to be excellent and assigned it a +9, then their average of +7 could be taken to be the evaluation of salary overall for this job and inserted into the *LMDF* as usual.

This remedy, however, turns out to be seriously flawed, and will lead to incorrect decisions in many cases. To see this, suppose that the $50,000 salary is almost certain to occur, and that the $35,000 salary, though still possible, is highly unlikely. Does it seem appropriate, then, that the overall evaluation of salary for this job should be exactly the average +7 of the evaluations of the separate salaries, precisely halfway between +5 and +9? Thinking about this, we realize that in fact it isn't appropriate, that the true evaluation should be better than that, much closer to +9 because of the virtual certainty of that larger salary. Using the average of +7, therefore, will almost certainly lead to an incorrect score for this job, and ultimately an incorrect decision.

The correct remedy, it turns out, is not to use a simple average of the separate evaluations but a *weighted* average—where the weights are exactly the *likelihood of each salary occurring*. Let's suppose, for instance, that the likelihood of the higher salary is 90%, and that of the lower salary is

10%. The resulting weighted average of the two evaluations is thus:

$$9(.9) \ + \ 5(.1) \ = \ 8.6$$

If one thinks about it, this is exactly the right value for the overall evaluation of salary for this job. Indeed, the separate evaluations of the two salaries are now weighted in exactly the right proportion according to their likelihood of occurring, with the more likely salary being weighted more heavily than the less likely one. Consequently, the overall evaluation is much closer to +9 than +5—as it should be. Only if the salaries happened to have an *equal* likelihood of occurring would a simple average of their evaluations—exactly halfway between—be appropriate. This same remedy, incidentally, would work equally well for any number of salary possibilities and any combination of likelihoods of their occurring; in such cases the weighted average would just contain a correspondingly larger number of terms. In any case, we have now obtained a perfectly suitable single numerical evaluation of salary for this job, and it can be plugged into the *LMDF*, and the decision process continued as usual.

The weighted average technique just described turns out to work equally well for obtaining suitable evaluations in many situations involving the same kind of uncertainty, and it gives rise, therefore, to an alternative decision model—a *probabilistic* model—which can be employed successfully for many decisions involving that kind of uncertainty. We summarize its steps, which are essentially the same as the original model except for the more complicated evaluation step involving probability-weighted averages. First, one chooses metrics exactly as before. Next, one assigns weights to these metrics, as before. Then one evaluates each option with respect to these metrics—but now in two possible ways. First, if an evaluation involves a single, certain outcome, then one performs the evaluation exactly as before. However, if an evaluation involves multiple possible outcomes as in the example above, then one proceeds as described there. Namely, one first evaluates each outcome as usual, then one determines their probabilities, and finally one computes the appropriate probability-weighted average of these evaluations—which becomes the single numerical evaluation with respect to that metric. Finally, once all such evaluations are complete, the scores for each option are computed as usual using the *LMDF*, and a decision can be made.

The upside of this alternative model is that it is considerably more general than the original and hence can be applied to many more decisions, particularly those troublesome and not-too-uncommon decisions involving uncertainty of the kind described. Its downside—and the price one pays for this greater generality—is that it is considerably more sophisticated and hence substantially more effortful to implement.

This model, broadly applicable as it is, does not apply to *all* decisions. First, it applies only to decisions involving outcomes whose probabilities can be determined. Second, like the original model, it does not apply to decisions involving off-the-scale weights or evaluations. Finally, it does not apply to decisions involving outcomes that are extremely rare—say with a probability much less than one percent—since in such cases the resulting scores can be so skewed as to produce absurd decisions. Such decisions must be handled by other means.

Decisions with Greater Uncertainty

Lastly, we take a look at a model that can handle decisions involving even greater uncertainty than the previous model, namely, decisions whose alternatives not only admit multiple possible outcomes, but outcomes *whose probabilities cannot be determined.*

To illustrate, we return to the example from the previous section where we supposed we were deciding between two jobs. Here, however, we assume more profound uncertainty, namely, only that *Job 1* offers between $40,000 and $60,000 annually, and *Job 2* between $50,000 and $55,000. Moreover, we assume that we don't know how likely any of these possible salaries are, which of course renders it impossible to employ the model of the previous section.

Not all hope is lost, however. It turns out that if we can obtain just two meager pieces of information about the salaries—namely their *best* and *worst cases*—then we'll still be able to use the *LMDF* to make some kind of useful decision. Fortunately, such information is easily obtainable for this example, and therefore we can proceed with a decision.

First we identify the best and worst possible salaries for each job, which are obviously $60,000 and $40,000 for *Job 1*, and $55,000 and $50,000 for *Job 2*. Next we perform *separate* evaluations of these salaries in the usual manner. This gives us a pair of evaluations for each job for their salary. For example, suppose these evaluations for *Job 1* are +8 and +4, and those for *Job 2* are +7 and +6. Next, instead of computing probability-weighted averages of these pairs as in the previous section in order to obtain a single evaluation for salary for each job—which cannot be done now anyway because of the lack of the necessary probabilities—we simply utilize *both* evaluations to compute *two separate* overall scores for each job. One of these scores is computed by plugging the best-salary evaluation into the *LMDF* along with any other evaluations that might be required, and the other is computed by plugging in the worst-salary evaluation along with these other evaluations. In the end, each job will have two scores, one representing its overall value with its best salary, and the other its overall value with its worst. For the sake of illustration, suppose we've done all this, and that these overall scores have come out to be +9 and +5 for *Job 1*, and +7 and +6 for *Job 2*.

An eminently useful decision can now be made. We note, however, that there is no hope for making the best possible decision and identifying the *definitively* better job, as there is simply not enough information. We know, for instance, that the best-salary score of *Job 1* is higher than that of *Job 2*—which might indicate that *Job 1* is the better job. But we also know that the worst-salary score of *Job 2* is higher than that of *Job 1*—which would seem to indicate the exact opposite! All other attempts to use the given information to identify a definitively better job likewise fail. Now, if it happened that *both* the best- and worst-salary scores of one job were higher than both such scores for the other, then there *would* be a definitively better job—the former—but this is unfortunately not the case for our example.

What kind of decision can then be made? It turns out that a variety of decision strategies have been formulated over the years to make useful decisions in the presence of such profound uncertainty. None identifies the definitively best option, but each identifies a preferred option in some specific sense. Here are three simple and fundamental examples of such strategies:

- *Optimize Downside:*Here we compare only the downsides of options and ignore their upsides, by choosing the option with the *best possible downside*, that is, the option with the *highest worst-case score*. This is a safe, conservative strategy that offers the best downside protection but leaves the upside to random chance. Applied to our example, this strategy identifies *Job 2* as the better job since its worst-salary score of +6 is better than the corresponding +5 score for *Job 1*. This decision ensures we never earn less than $50,000, but it also ensures that we never earn more than $55,000.

- *Optimize Upside:* Here we compare only the upsides of options and ignore their downsides—the opposite of the preceding strategy—by choosing the option with the *best possible upside*, that is, the option with the *highest best-case score*. This is a risky strategy that offers the possibility of maximizing the upside but leaves the downside to random chance. Applied to our example, this strategy identifies *Job 1* as the better job since its best-salary score of +9 is better than the corresponding +7 score for *Job 2*. This decision gives us a chance to earn $60,000, but it also leaves open the possibility of earning only $40,000.

- *Optimize Average of Upside & Downside*: Finally, there is a simple way to take into account *both* the upsides and downsides of options—by choosing the option with the highest *average* of its best- and worst-case scores. What this means for two options, in effect, is that if there is no definitively better option having both upside and downside gain over the other option, then this strategy will choose the option with greater upside gain than downside loss, or greater downside gain than upside loss. It is a more balanced strategy than the two extreme strategies above, and represents a sort of compromise between them. Applied to our example, this strategy identifies *Job 1* as the better job

> since the average +7 of its best- and worst-salary scores is better than the corresponding average +6.5 of *Job 2*. Indeed, *Job 1* has more upside gain than downside loss, since its best-salary score is 2 points higher than that of *Job 2*, but its worst-salary score is only 1 point lower.

A few remarks about these strategies are in order. First, they will generally produce very different decisions—as with our example—because each identifies the better option in a very specific sense. The happy exception are decisions for which it happens that there *is* a definitively better option—when both the best- and worst-case scores of one option are higher than both such scores for the other—in which case all three strategies will correctly identify it. Finally, regarding which strategy to employ, there is no simple answer—one must simply use one's best judgment as to which strategy is most appropriate for a given decision.

The decision model just outlined for the example works equally well for many other decisions involving the same kind of uncertainty. We summarize its steps. First one chooses metrics exactly as before. Next, one assigns weights to these metrics, as before. Then one evaluates each option with respect to these metrics—but now in two possible ways. First, if an evaluation involves a single, certain outcome, then one performs the evaluation exactly as before. However, if an evaluation involves multiple possible outcomes whose likelihoods are not known, as in the example above, then one proceeds as described there. Namely, one first determines the best- and worst-case outcomes, and then one evaluates both these outcomes in the usual manner to obtain a pair of values. Once all these evaluations are complete, the next step is to compute scores for the options using the *LMDF*—for which there are now two possibilities. First, the options involving no uncertainty will yield single scores as usual. However, the options involving uncertainty will yield pairs of scores—their best- and worst-case scores—the former resulting from the *LMDF* with all the best-case evaluations plugged in, and the latter resulting from all the worst-case evaluations being plugged in. Finally, with all these scores in hand, a decision can be made using one of the three strategies described above.

The good news about this alternative model is that it is very general and hence can be applied to many more decisions than both the original and probabilistic models, particularly those troublesome decisions involving such profound uncertainty that not even the probabilities of outcomes are known. The only decisions to which it does not apply—besides those for which the necessary best- and worst-case scenarios cannot be determined—are those decisions, like with the original model, that involve off-the-scale weights or evaluations. Further good news about this model is that it is not difficult to implement—it is only slightly more complicated than the original and considerably simpler than the probabilistic model.

The bad news about this model—and the price one pays for its simplicity and great generality—is that its decisions are often far inferior to those produced by the previous models. In the presence of such profound uncertainty, however, such decisions are the best one can usually hope for, and are still far superior to random luck.

72226638R00067

Made in the USA
Columbia, SC
14 June 2017